FROM THAT TO THIS

From That to This is published under Erudition, a sectionalized division under Di Angelo Publications, Inc.

Erudition

Erudition is an imprint of Di Angelo Publications.
Copyright 2023.
All rights reserved.
Printed in the United States of America.

Di Angelo Publications
4265 San Felipe #1100
Houston, Texas 77027

Library of Congress
From That to This
ISBN: 978-1-955690-28-7
Paperback

Words: Nemat Saidi Lavasani
Cover Design: Savina Deianova
Interior Design: Kimberly James
Editors: Cody Wootton, Willy Rowberry

Downloadable via Kindle, NOOK, iBooks, and Google Play.

For educational, business, and bulk orders, contact distribution@diangelopublications.com.

1. Science --- Study and Teaching
2. Science --- Space Science --- General
3. Science --- Physics --- Optics & Light
4. Medical --- Radiology, Radiotherapy and Nuclear Medicine

FROM THAT TO THIS

NEMAT SAIDI LAVASANI

CONTENTS

Benam-e-khoda
(in the name of God)

PREFACE

13.8 billion years ago, the big bang was the conception of life in the universe, and 9.2 billion years after that, the first thermonuclear fusion in the heart of our sun was the birth of humanity itself.

From That to This is a journey through the evolution of cosmic gases and dust, into the simple element of hydrogen, and later into heavier elements such as uranium, neptunium, and plutonium. But our nuclear physicists did not stop there and made some twenty more heavier elements in their laboratories, which now are part of the periodic table of elements—the periodic table of elements that was invented by Dmitri Mendeleev in 1869. Then, it only contained ten elements and four predicted ones.

However, the whole spectrum of electromagnetic (EM) radiations has been practically unchanged since the big bang. As a matter of fact, cosmic microwave background radiations (CMBR) still linger around the whole universe from the time of the big bang, from which astrophysicists can measure and calculate the authenticity of the big bang theory.

This book will take you through the macro world of stars, galaxies, and how our astronomers build bigger and better telescopes to look deeper and deeper into heaven, while our nuclear physicists delve into the micro world of quantum mechanics.

Quantum physics was practically started in 1865 by James Clerk Maxwell, who formulated the theory of electromagnetic radiation. Thereafter, the discoveries of X-ray, radioactivity, and other constituents of the atom, such as alpha, beta, gamma, electrons, protons, and neutrons, opened the floodgate to the discoveries of the elementary particles to finally formulate the Standard Model of the elementary particles. Nowadays, the Standard Model of elementary particles is the bible of quantum physics.

In this book, I also talk about the creation of the atom bomb and its unimaginable destructive power, and later, through the end of WWII, how the scientists and the politicians came to their senses and dedicated the whole knowhow of the atomic industry to humanitarian efforts— such as lighting up thousands of cities, major environmental protection, and saving millions of lives in the medical field all around the world.

In the final chapter of the book, there is an emphasis on some of the procedures done in the Department of Nuclear Medicine in order to familiarize college students and some patients who may need some clarification on these procedures. I am sure the general public can benefit from this information as well.

And lastly, I apologize for any shortcomings in this book and welcome any constructive criticism moving forward. God bless!

MY QUALIFICATIONS

I have a Master of Public Administration degree from California State University, Dominguez Hills.

I worked as a nuclear medicine technologist for over thirty years at UCLA Medical Center and Dignity Health - California Hospital Medical Center in Los Angeles.

I have spent the last four years of my retirement researching the path of radiation from the time of the big bang to today's use of radiation in wireless communications and in departments of radiology in hospitals and other industries.

CHAPTER ONE

"If you know not to which port you sail, no wind is
favorable."
—Lucius Annaeus Seneca

MASS MATTERS

Not long ago, I was watching the World Deadlift
Championships on TV when I realized something
profound: I was witnessing a miniature *big bang* happening
right before my eyes.

One of the competitors, Miguel Siblicov, arrived at the
arena, waving his hands in the air in response to his fans.
There was a lot of noise—the type that creates a feeling of
anticipation in you, that makes you know you are about to
witness something extraordinary.

Miguel stood behind the bar of a huge weight that the
electronic board read as weighing in at 426 kg, equal to
937 lbs.

He looked up as if in prayer and slowly took in a long, deep breath. Then he bent down and grabbed the bar. Miguel gingerly turned his hands back and forth halfway around the bar before carefully tightening the leather straps around it. He then took a firm hold with both hands, delicately adjusted his feet under the bar, slowly lowered his butt, and, ever so cautiously, started to straighten his back . . . A moment of quiet fell.

He let out a loud scream as if to awaken all the sleepy cells in his body to help him in his endeavor!

Coaches clamored and fans methodically chanted, "Lift! Lift! Lift!"

Miguel released another guttural groan during the liftoff and launched the heavy weight four-fifths of the way up—but there was still another one-fifth to go. The weights on both sides of the bar teetered and shook wildly, almost as if they were stubbornly refusing to be defeated by Miguel.

Simultaneously, the whole arena seemed to be on steroids, screaming, "Up! Up! Up!" which encouraged Miguel not to give up. No, he wasn't about to disappoint his fans. Not Miguel! He was pressing on. It was amazing to watch! You could see the fight in his red face, glistering with sweat. He was still fighting back as he pulled the weight, millimeter by millimeter, against his legs.

Then, among all the commotion and uproar, the piercing cry of a siren could be heard as the officials called out, "It was a good lift!"

The whole arena broke into a loud cheer, celebrating Miguel's victory. Some people were high-fiving, hugging each other, and laughing.

Amid all that pandemonium, something unusual happened: a gush of blood streamed from Miguel's nostrils. Miguel dropped the weights, triumphantly waved to his fans, and smiled ear to ear as he walked away with his coaches in good spirits.

So, what had just happened? It was a microscopic, miniature version of the big bang! In later chapters, we will read about the interesting life and death of all the stars. The processes of implosion and explosion of the stars are yet another demonstration of a miniature big bang.

In Miguel's case, imagine that his entire body was the core of an *accretion disk*. An accretion disk is a galaxy-sized disk composed of accumulated interstellar media (dust and gases). When the accumulation reaches a certain point, the gravitational force and pressure at the core reaches a critical point that is ripe for the processes of a *nuclear fusion reaction,* and that is the moment that the *explosion* will take place. The explosion that took place at the beginning of the universe is called the big bang. In Miguel's case, his was a bloody nose!

As the extra weight came upon Miguel's strained body, his internal resistance could hold on for only so long. His red, sweaty face was a testimony to the magnitude of his body's resistance, but at one point, his internal resistance was overcome by the external weight's pressure and heat, causing the rupture of the capillary veins inside his nostrils.

Although a simple occurrence in comparison, a few milliliters of blood pouring out of Miguel's nose mimicked the insanely complex processes of what is referred to as "the big bang." The big bang explosion was an unimaginable amount of hot sizzling *plasma* thrown into the space that became the seeds of all the galaxies in the universe! And that event took place 13.8 billion years ago.

Right after the big bang, *cosmic inflation* immediately occurred in a nanosecond. Following that, it took a period of 100 to 200 million years to allow for the dense and hot hydrogen plasma and radiation to sprawl out and cool down. Thereafter, cosmic inflation provided an ideal environment for protons and electrons to slow down and form hydrogen atoms, which, back then, constituted 99.99% of the universe.

Notably, the big bang is our most modern explanation of the origin of the universe—the subject of an age-old debate. There's been a back-and-forth argument laced with rhetoric among theologians and scientists alike—and they have yet to come to a mutual understanding of the subject matter!

The reason it has taken such an unimaginably long time to come up with a reasonable agreement is simply this: there exists a huge discrepancy in opinions. On one hand, there are those who wholeheartedly believe in the preordained six holy days of the creation of the world by God Almighty. And on the other side, there are the scientists who believe data suggesting the universe took 13.8 billion years of evolution to create by the power of Mother Nature herself.

However, some prominent scientists have a different view of God. *Paul Dirac,* one of the greatest mathematicians of the twentieth century, said, "God used beautiful mathematics in creating the world."

Albert Einstein was agnostic and still believed God could

REDSHIFT SPECTRUM

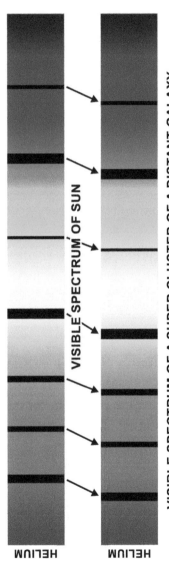

RED SHIFT

VISIBLE SPECTRUM OF A SUPER CLUSTER OF A DISTANT GALAXY

AND THE OPPOSITE SHIFT WOULD BE "BLUE SHIFT".
"**SPECTRAL LINES**" ARE THE FINGERPRINTS OF ELEMENTS.
EACH SPECIFIC SET OF BLACK LINES IS REPRESENTING A SPECIFIC ELEMENT.
IN THIS CASE, SPECTRAL LINES OF HELIUM ARE SHOWN.

Fig. 1

have been involved in the world's creation. He said, "I want to know how God created this world. I want to know His thoughts; the rest are details."

Now, do you think proponents of these two opposing philosophies can sit at a table, argue their cases, and come up with an agreement somewhere in the middle? I strongly doubt it! There is too great a discrepancy between the foundations of their thoughts.

FIVE REVELATIONS OF PROVING THE BIG BANG THEORY

To follow the scientific road (because the theologian's belief is set and there are no ifs, ands, or buts about it), I identified *five chronological revelations* that summarize what hundreds of theoretical and experimental physicists, astronomers, mathematicians, and scientists have worked on for hundreds of years.

Essentially, these five revelations encapsulate the endlessly long hours of arduous work spent researching the big bang by multitudes of hardworking scientists, some of whom have been Nobel laureates for their wondrous contribution to this field. In short, these five revelations convinced the scientific world that the theory of the big bang is, in fact, valid and correct! At least, for the time being.

REVELATION ONE

In 1915, Albert Einstein's *theory of general relativity* opened the door to a better understanding of gravity and its interactions among all matter in the universe. It was applied to *redshift* and the *receding of galaxies.*

Einstein said that space and time are not separate, but they are of the same entity, and he called that entity *space-time*. Furthermore, he explained that matter and radiation curve, bend, and disturb space-time. In turn, *curved space-time* dictates the movements of the matter and radiation, which Einstein called *gravity*.

REVELATION TWO

In 1917, American astronomer *Vesto Slipher* experimentally demonstrated the phenomenon of redshift.

Slipher compared the spectrum of the sun's visible light to the spectrum of the light coming from a distant galaxy (Fig.1).

Slipher demonstrated that the *Doppler effect*—that sound waves coming from an approaching train gain frequency, becoming higher in pitch, and sound waves from a train speeding away stretch out, becoming lower in pitch—also applies to *light waves* from stars in the universe.

Light waves will shift to higher or lower frequencies according to the movement of the light source. The speed of light never changes and is always constant; therefore, if the light source is moving away from the observer, then the *frequencies* of the light will be stretched out like an accordion, causing a *longer wavelength* and thus *lower frequencies* (*redshift*). And vice versa, when the light source is coming toward an observer, then the frequencies of the light will be condensed, causing a *shorter wavelength* but *higher frequencies* (*blueshift*).

Let's look at the small band of the *visible light* within the electromagnetic (EM) spectrum. Red is at the starting point, with the lowest frequencies and longest wavelength, versus the blue light, with the highest frequencies and shortest wavelength at the end of the visible light band.

Slipher's redshift experiment proved that galaxies are receding, or moving away from us. Our universe is expanding.

REVELATION THREE

In 1927, *Georges Lemaître,* a Belgian Catholic priest, mathematician, astronomer, and physicist who worked with American scientists in the field of astronomy, also validated and supported the theory of the expanding universe. Lemaître claimed that if the universe is currently expanding, then in the past, the universe must have been smaller. He traced it back to an originating single point, which he called the *primeval atom.*

One of the opponents of Lemaître's theory was the British astronomer, Fred Hoyle. He coined the name "the big bang" to mock Lemaître's theory. Ironically, the name still stands.

REVELATION FOUR

In 1929, American astronomer *Edwin Hubble* experimentally discovered that many celestial objects, previously thought to be heavy clouds of dust and gas, classified as *nebulae* outside of our Milky Way Galaxy, were in fact galaxies.

Additionally, his observational *expansion of the universe theory* served as one of the key pieces of evidence most often cited in support of Georges Lemaître's big bang model.

The final revelation that locked down the seal of approval onto the big bang theory—and made concrete the notion of an expanding universe—was the accidental discovery of cosmic microwave background radiation.

In 1964, American radio astronomer *Robert Woodrow Wilson* and American physicist *Arno Allan Penzias* were employed by Bell Telephone Laboratories to construct a radio receiver. Working on their project, they took advantage of an obsolete satellite antenna called the Holmdel Horn Antenna (which, in 1990, became a United States National Historic Landmark). While working with it, they noticed a very odd buzzing sound. They started to investigate by eliminating all the possible sources of the annoying signal. Eventually, they discovered that the annoying buzzing sound was strangely coming from every direction . . . and at any time of day or night. They were perplexed! They could not come up with an answer.

Luckily for them, at the same time, another team of astronomers and physicists headed by *Robert Dicke* at the University of Princeton were trying to find the *thermal echo* of the universe's explosive birth. Basically, they were looking for the same thing that the two annoyed scientists at the Bell laboratory had accidentally stumbled upon. Serendipitously, one's problem turned out to be the other's solution.

Later, this phenomenon was called *cosmic microwave background radiation*. Wilson and Penzias both went on to receive the Nobel Prize in physics for this discovery in 1978.

So, what is CMBR? Right after the big bang, the universe was very hot and quite dense. But thereafter, during the cosmic inflation, the universe started to expand and cool down, providing the right conditions for the formation of subatomic particles and the subsequent formation of the galaxies!

Simultaneous to the cooling down of the hot plasma in the universe, the background radiation was getting cooler and cooler as well. The background radiation is the leftover material from the initial temperature. Today's CMBR is the remnant of that intensely hot period of the big bang.

Today's CMBR temperature has dwindled to 2.7 K (kelvins) above absolute zero (absolute zero = 0 K; -460 °F; -273 °C).

After hundreds of years of persistent work by many determined human thinkers, finally, in 1964, scientists came up with the concrete evidence that would verify everything.

A SHORT REVIEW OF THIS SECTION

First, the whole observable universe is expanding. *Second*, the expansion must have had a single starting point (singularity / the primeval atom) from which the extreme pressure and heat caused the big bang. *Third*, the residual heat (2.7 K CMB) lingering around the universe originated from the big bang.

Eventually, after the scientists were satisfied with their discoveries about the big bang and all of the high-fiving and hugging were over, some of the robo-scientists started to scratch the backs of their necks, squint their eyes, and show their teeth, finally presenting their colleagues with this question: If our universe is expanding, will it eventually slow down, stop, and start coming back?

If it does, would it be a hypothetical scenario called the *Big Crunch,* the start of an implosion of the universe (blueshift), going back to a singularity point, or the Georges Lemaître's primeval atom—a point too dense and too heated—which, in turn, could cause the next big bang?

For those who are losing sleep over the subject of the *Big Crunch*, scientists tell us not to worry! In five billion years, our sun will run out of hydrogen fuel at its core, which will be the end of its thermonuclear fusion processes; that will practically be the beginning of the end of our sun's life. It will enter the cycle of a *red giant* before eventually consuming three of the four inner planets—Mercury, Venus, and Earth—and ending life in our solar system as we know it.

Second, our Milky Way Galaxy is in a binary situation with another galaxy of similar size called the *Andromeda Galaxy.* The two galaxies are approaching each other with a speed of seventy miles per second (MPS). Eventually, in about five billion years, they will collide and create a new galaxy twice as big, which astronomers have a name for it already. It will be a giant elliptical galaxy named "Milkdromeda."

Third, the findings show that only 4% of the universe is composed of visible atoms, matter, or galaxies, while 26% is dark matter, and the remaining 70% is dark energy, meaning there is a lot of room to roam around.

STELLAR CLASSIFICATION SYSTEM

After the big bang—a period of 100 to 200 million years—spewed plasma spread out into space, cooling down to the point that the ionized particles and gases clumped together, becoming numerous accretion disks, moving and spinning, forming the seeds of new galaxies. Every accretion disk became a galaxy with a *central bulge* that contained most of that galaxy's mass, containing a black hole at the center surrounded with billions of stars. On the periphery of the disk, there are less massive celestial bodies.

Galaxies come in four shapes: *spiral, elliptical, lenticular,* and *irregular*. The most common ones are spirals, and our Milky Way Galaxy is one of them.

Scientists believe that there are more than 700 billion galaxies in the universe, each containing approximately 400 billion stars, with each star having an average number of 1.5 planets. Each planet can have between no moons and tens of moons.

To classify all the stars in the universe, there are three main factors that matter the most: *first*, the *temperature* of the star; *second*, the *strength of the absorption lines* within each spectral type; and *third*, the *luminosity level* of each star.

The first spectral classification of stars came about during the late 1800s by the astronomers at Harvard College Observatory, and was called the "Harvard Spectral Classification of Stars." In this system of classification, *temperature* was the main factor of distinction between different classes of stars. They had devised seven classes of stars based on their surface temperature. Each class was assigned a letter: O, B, A, F, G, K, M. The hottest class was assigned the letter "O" while the letter "M" represented the coolest class. A fun way to remember the order of letters in the acronym is by singing, "Oh, Be A Fine Girl/Guy, Kiss Me."

However, in the early 1900s, several astronomers, such as Annie Jump Cannon, Hertzsprung–Russell (HR), and Morgan–Keenan (MK), noticed that there were other factors that must be considered for the stars' classifications.

Two more factors were added: the *strength of the absorption lines* within each spectral type, and the *luminosity level* of each star.

For each of these two added factors, there are *subclasses* that are shown by numerical values. For the strength of the absorption lines, the numerical values of zero to nine divide the subclasses. For the luminosity factor, the Roman numeral values of one through seven (I, II, III, IV, V, VI, and VII) show different subclasses of the star's luminosity level.

For example, our sun is classified as G2V. That means our sun is a G-type star in terms of temperature, with an absorption line strength of 2 and the luminosity value of Roman numeral number five (V).

So, the next time you look at a colorful picture of a galaxy, you can tell what type of stars there are in that galaxy just by noticing their colors. Blue stars are the most massive and hottest, then whitish-gray, yellow, orange, and red stars, from the most massive and hottest to least massive and coldest, respectively.

If you happen to be looking at a beautiful picture of the

Milky Way Galaxy, our solar system is located within one of the galaxy's *minor arms* called *Orion*. Our solar system is 28,000 light years from the center of our galaxy.

During antiquity, our thinkers surmised that our sun was the center of the universe; however, no one has since presented concrete evidence that our ancient thinkers were wrong!

In summary, our scientists have created standards that measure the mass and the luminosity of all other stars in the universe relative to the mass of our sun (M☉) and the luminosity of our sun (L☉). It matters not if the stars out there are millions or billions of times more massive, smaller, brighter, or dimmer than our sun.

SIZES OF THE STARS

To classify the stars in terms of their sizes, there are three groups to consider. *First,* there are *small stars,* such as red, white, brown, and black dwarfs. *Second,* there are *main sequence stars* or intermediate-mass stars, such as our sun. *Third,* there are *supermassive stars,* such as Spica in the constellation Virgo.

Main sequence stars comprise 90% of all stars in the universe. The reason for this is because the massive stars come and go much faster than the smaller and average-size stars, because their central nuclear combustion is much faster and stronger, causing them to burn themselves out in a shorter time.

In terms of small stars, since they live much longer than 13.8 billion years, and the medium-size main sequence stars' life spans are within 10 billion years, few of them have turned into long living small stars at this point in time.

In terms of stars' masses, stars' classifications do not necessarily mean that the bigger they are, the heavier they must be. Stars such as neutrons, quasars, and pulsars are much smaller than our sun, but they are up to ten times heavier. What is important about the classification of stars is that it is their mass which dictates the cycle of their *formation, life,* and *death*.

For example, our sun, being classified as a main sequence star, will live as a *yellow dwarf* star and die as a *red giant*, leaving behind a *planetary nebula* from its outer layers and a *white dwarf* star from its core.

Supermassive stars will die as *red supergiants,* and eventually, they will explode into a *supernova,* leaving behind a *neutron star*.

All stars arise from different sizes of nebulae (clouds of dust and gases). After a section of a nebula gains sufficient mass, it becomes a *protostar*, which collapses under its own gravity, triggering a *thermonuclear fusion* of hydrogen atoms into helium atoms. This nuclear reaction releases a huge amount of *energy* that pushes upward to stop further *gravitational force collapse*, and the surplus energy will heat and light up the newborn star. The first nuclear fusion at the core of any star is its first heartbeat, celebrating its birth into stardom.

The lifespan of every star is dependent upon their size. The smaller they are, the longer they live; the bigger they are, the faster they burn out and die.

The O-type stars 30 to 60 times the mass of our sun (MO) have a lifespan of 3 to 10 million years, while our sun's lifespan is 10 billion years, and the least massive stars of M-type can live even longer—about 1 to 10 trillion years. Astronomers believe that bigger stars produce much more pressure and heat at their core and burn their central fuel faster, eventually dying sooner.

The big stars play hard and die early.

When the stars of any mass class die, depending on their sizes, they leave behind different classes of stars.

An average-size star, such as our sun, at the end of its life, it will fall into a red giant cycle, during which their outer layer will spread into the space creating a *planetary nebula*. Then, the planetary nebula will turn into *interstellar medium* (dust and gases) and eventually belong to other galaxies. The only part left from the average-size stars (main sequence stars), including our sun, will be their core, called a *white dwarf*.

All massive stars—such as red supergiants and supernovae—will explode into space at the end of their lives, leaving behind two distinct masses: *first*, star-forming nebulae from their outer layer, and *second*, neutron / pulsar stars from their core.

Furthermore, it is fair to say that stars of any size are born and die in their cores. They are born when the first nuclear fusion takes place in their core, and they will die when the last nuclear fusion takes place in their core.

In the case of medium-sized / main sequence stars, when a star runs out of hydrogen atoms in its cores, the star's mass is not great enough to create enough heat to fuse heavier elements, and the star will die.

And in the case of super massive stars, the star's mass

is only great enough to create enough heat and energy to fuse the elements *below* the element iron (in the periodic table of elements), and is not able to fuse elements heavier than iron, so the nuclear fusion combustion will stop. As a result, there would be no energy to fight back the star's own downward gravitational force, so the upper layers will start to collapse and implode on their own core. The implosion will produce an internal shock that will blow up the star's outer layers.

Now, one might ask, if the heaviest element being produced by any supermassive star is iron, then where do the other heavier elements we have here on Earth come from?

The elements heavier than iron are not created in the belly of any supermassive stars, but they are the product of the collisions of stars such as neutron stars, black holes, quasars, or pulsars.

When these supermassive stars fall into a deadly dance of tango, it's called a *binary* situation, which makes the collision of the two stars inevitable. Their collisions are powerful enough to create enough heat and energy to cause fusion of iron atoms together, creating heavier elements in the periodic table of elements, but only up to the element of U-92.

Since we have the heavy element of plutonium here on Earth, then one might ask, how did they get here? The only way this could have happened is if the *solar nebula* that formed our *solar system* was formed from the collision of binary supergiant stars, providing us with these heavy elements we have here on Earth.

Furthermore, if the element uranium is the heaviest element naturally existing on Earth, where do the rest of the heavier elements in our *periodic table of elements* come from? All the heavy elements after uranium are made here on Earth in our laboratories, nuclear reactors, and particle accelerators, by our own mighty nuclear physicists.

As we see, there are 118 elements in our periodic table of elements—the last one is the element oganesson (Og-118), with 118 protons in its nucleus (Fig. 12).

OUR GALACTIC ADDRESS

When we talk about distances within the solar system, we are using a yardstick called *astronomical unit* (au), which is the distance between the Earth and the sun—93 million miles, or about 8 light minutes.

However, when we step outside of our solar system, a bigger measuring stick is needed. The galactic measuring stick is called a *parsec* (pc).

WHAT IS ONE PARSEC?

Fig. 2 shows how astronomers came up with this new measuring stick called parsec. What was needed for the calculation of one parsec was an imaginary right triangle, the distance between the sun and Earth (one AU), finding the nearest star to the sun (Proxima Centauri), and two observations six months apart, to find the *apparent parallax motion* of the Proxima Centauri. Referring to Fig. 2, we can see that when the parallax angle is equal to one *arcsecond* (explained below), then the distance from that point to the center of the sun will be one parsec.

One parsec is equal to 19.2 trillion miles, or 3.26 light years (ly), or 210,000 au.

What does "one arcsecond" mean?

In a circle, one full arc is 360 degrees;

then 360x60=21,600 arcminutes,

and one arcsecond = 21,600x60=1,296,000.

So one arc second is equal to 1/1,296,000 of one full arc.

Where did the name "parsec" come from? It is a name taken from "**par**allax angle" and "arc**sec**ond."

Then, knowing the two measuring sticks of "au" and "pc" enables our astronomers to establish a map of the universe. A bird's eye view of our visible universe from the Hubble Space Observatory and other sources tell us that there are more than 10 million *superclusters*. They are the largest known structures in the visible universe.

Each supercluster is about 100 MPC across and they all are separated from one another by a huge space called *void*. Our galaxy is part of the supercluster called the *Virgo / Local Supercluster*.

The next level below the superclusters is the *galaxy clusters,* which have a 33-MPC (110-Mly) diameter and a mass of 10^{14} to 10^{15} MO. Each contains 100 to 1,000 galaxies. We are a part of the galaxy cluster called the *Virgo Cluster*.

Each galaxy cluster contains many *galaxy groups*, and each galaxy group is made of 54 galaxies or less.

We are part of the galaxy group called the *Local Group*. Two of the biggest galaxies within the Local Group are located at the center of the cluster; named the *Andromeda Galaxy* and the *Milky Way Galaxy*, they are only separated

PARSEC

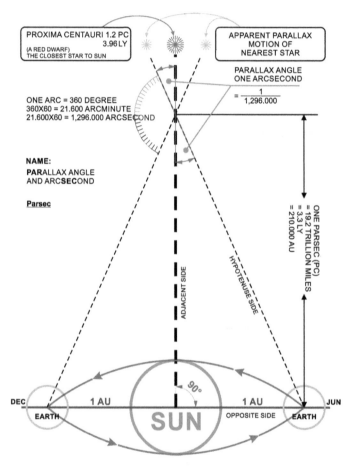

Fig. 2

by 0.8 MPC (2.6 Mly) of distance from each other. These two galaxies are locked also into a deadly dance called a *binary situation*, spiraling closer to each other with a velocity of 70 miles per second or 113 km/s. Their collision is predicted to occur in about 5 billion years from now.

But unfortunately, they don't know if it is going to happen in the morning or in the afternoon. :)

(The latter piece was a part of P. Reagan's joke.)

Astronomers predict that when the Andromeda and the Milky Way galaxies collide, the new galaxy is going to be a giant elliptical galaxy, already nicknamed "Milk-omeda" or "Milk-dromeda."

The two galaxies of the Andromeda and the Milky Way are described as dumbbell-shaped because they both possess almost the same amount of mass, each one approximately one trillion times the mass of our sun (10^{12} M☉), and both of these two galaxies are beautiful, round spiral-type galaxies (Fig. 3).

OUR UNIVERSAL ADRESS

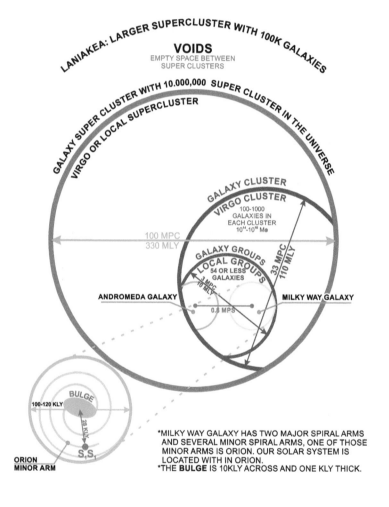

Fig. 3

CHAPTER TWO

"God is a mathematician of a very high order.
He used advanced mathematics in constructing the
universe."
—Paul Dirac

MILKY WAY GALAXY

Outside of our Milky Way Galaxy, distances are measured in *millions* of parsecs, while within the galaxy, distances are measured in *thousands* of parsecs and/or light years.

The Milky Way Galaxy is categorized as a *spiral-shaped galaxy* with four spiraling tentacles, called the *arms*. Some astronomers believe there are only two *major arms* and several *minor arms*. *Orion* is one of those minor arms, and it is home to our solar system. Our solar system is 28 thousand light years from the galaxy's black hole.

At the center of the Milky Way Galaxy, there is a disk-like *bulge* compacted with stars, comprising 10 to 20% of the mass of the whole galaxy. The bulge is 10,000 ly wide and 1,000 ly thick.

At the center of the Milky Way Galaxy, *Sagittarius A** and a supermassive black hole are lurking. The supermassive black hole is an intense *radio source,* and it is about 4 million times the mass of the sun (MⴰO). The Milky Way Galaxy is 100 to 120 Kly in diameter.

Beyond the Milky Way Galaxy, there is a halo of *dark matter* 2 million light years across.

The Milky Way Galaxy houses approximately 100 to 400 billion stars, and 600 billion planets.

SOLAR SYSTEM

Roughly 9.2 billion years after the formation of our Milky Way Galaxy, our solar system was formed. Astronomers believe that our solar system must have been formed from a *nebula* called the *solar nebula,* a nebula which had been left behind from the explosion of two *supernovae.* They believe this because, here on Earth, we have such a heavy element as uranium (U-92), which can only be produced from the collision of supermassive stars.

The solar nebula's molecules, like any other nebula's molecules, were not uniformly dispersed, and, in some areas, they were denser than others. The densest and most massive parts of a nebula are called *protostars*, which can later form into stars. (In our solar system, our sun took 99.98% of the mass of the whole solar nebula, and only the remaining .02% went to eight planets and the rest of the objects within the solar system.) Outside of the protostar cloud, the less dense areas were called *protoplanets*, which later formed all the planets and smaller bodies.

SOLAR SYSTEM

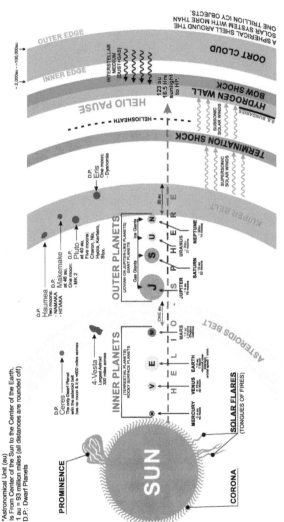

*Astronomical Unit (au)
Is From Center of the Sun to the Center of the Earth.
1 au = 93 million miles (all distances are rounded off)
D.P.: Dwarf Planets

Fig. 4

To start the processes of our solar system, the solar nebula or the *molecular clouds* generally must receive an outside shock, either coming from a *passing star* or a shockwave from a *supernova,* which will trigger and cause the collapse of protostars and protoplanets onto their own core under their own gravity. In the case of *star formation*, a star must be massive enough to create at least 3.8×10^{12} psi (pounds per square inch) of gravitational force / pressure at its center, which, in turn, will generate 15 million °C (27 million °F) of heat to ignite the processes of thermonuclear fusion. Then, the processes of hydrogen fusion will provide the star with enough energy to stabilize the star's condition to fight back the gravitational force; also, there will be a surplus of energy that will start penetrating through the upper layers of the star and eventually show up at the star's surface as heat and light.

The moment that the first thermonuclear fusion takes place at the core, it will be the inauguration of a protostar into the stardom!

After the formation of the star, all the leftover materials, called *protoplanets,* will form the upcoming planets and smaller celestial bodies, such as dwarf planets, asteroids, comets, meteorites, and *Kuiper belt objects (KBO).*

In our solar system, the planet formation processes took place another 100 to 200 million years to cool down, during which the smaller planetary bodies collided with one another over and over to gain more mass, forming the planets that they are today.

For a planet to be recognized as a planet, and not to be confused with a dwarf planet or other smaller bodies, there are *three* important criteria that must be met.

First, planets must have their own exclusive orbits with a clear path around the sun.

Second, they must orbit the sun.

Third, the planets must have a spherical shape.

Dwarf planets and other non-planets lack one or more of these three criteria.

In 2006, planet *Pluto* was reclassified as a *dwarf planet* because it did not have its own clear orbital path going around the sun, although it is spherical and orbits around the sun. Pluto is orbiting the sun among millions of other KBO.

After the planets were formed, the positioning of all planetary bodies around the sun was the work of the *solar wind*, which pushed them away from the sun in the order they are in today. The first four *rocky planets* are called the *inner planets*, which are closest to the sun (Mercury, Venus, Earth, and Mars). The last four planets are called the *outer planets* or the *giant planets;* the outer planets are made up of two *gaseous planets* (Jupiter and Saturn), and two *icy planets* (Uranus and Neptune). The planets Uranus and Neptune, and all the KBO, are frozen because they are so far away from the sun (Fig. 4).

SUN

Around 13.8 billion years ago, the Big bang was the conception of life in the universe, and 9.2 billion years after the Big bang, the first spark of thermonuclear fusion in the heart of our sun was the birth of humanity as we know it today.

Our sun is 4.6 billion years old, and it will shine for another 5 billion years before falling into a disastrous, deadly cycle called the *red giant cycle*. The red giant cycle awaits all main sequence stars at the end of their lives.

At this point, there are two questions that must be answered.

First, what are main sequence stars?

Second, what is a red giant cycle?

First, main sequence stars are a group of stars within the "G" classification of stars, like our sun. The masses of these stars range somewhere between ½ to 10 M☉ (M☉= $2x10^{30}$ kilograms), and their average lifespan is around 10 billion years.

In general, the supermassive stars shine brighter and hotter and will die much younger, with an average life span of 10 million years. The "M" type small-size stars will live for an average of 10 trillion years.

Second, the red giant cycle will happen to all main sequence stars at the end of their lives. Stars in general exist because they possess enough *mass* to create the *pressure* and *heat* needed for the fusion of *hydrogen atoms* in their core, thus creating enough energy to keep the star stable, hot, and shiny. However, when the supply of hydrogen atoms is exhausted, the production of thermonuclear force will cease and the star's upper layers will start crumbling down onto its core. The implosion causes a huge amount of heat, which, if it is intense enough (around 100 million Kelvin), will cause the helium atoms that are the byproduct of the hydrogen fusions to start fusing together through a reaction called the *triple-alpha process,* converting three helium nuclei into one carbon atom. The *helium fusion* will produce a great amount of thermonuclear force, keeping the star alive for a relatively short time until the helium supply runs out as well. At that time, the star will implode, then explode and start to swell.

And that is the start of a red giant cycle.

In the case of our sun, it will swell up more than one AU, engulfing three of the four inner planets—Mercury, Venus, and our beautiful Earth—and ending life in our solar system as we know it today!

The cycle of red giant takes about 10 million years, during which the main sequence star will be divided into two distinct celestial bodies: *First*, the outer layers that will become a *planetary nebula*, drifting away in the space. *Second*, the core of the star will remain as a *white dwarf*. Since there are no nuclear fusions taking place at the cores of these white dwarfs, they will gradually lose heat and luminosity, turning into brown and then black dwarfs that will go on living for trillions of years.

Throughout history, our sun has been worshipped and admired by all humanity for its beauty, light, heat, fertility, and above all, its mighty power. However, at the present time, the way we worship it is different than the way early man did. The modern man has taken the sun out of its piety category and classified it as only a yellow dwarf G-type star. Now we know there are some stars 300 times bigger and 10 million times brighter than our sun in another galaxies, 165,000 light years away. By learning more about our sun, we have come to the conclusion that our sun is

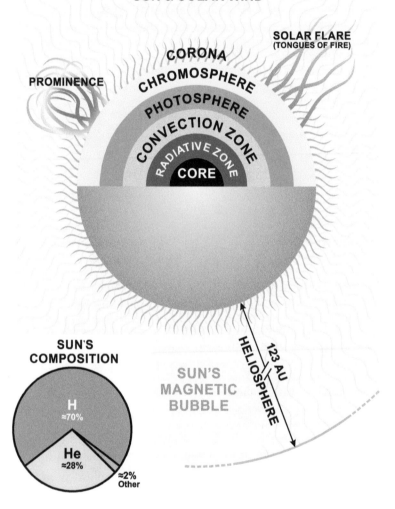

SUN & SOLAR WIND

SOLAR FLARE
(TONGUES OF FIRE)

CORONA

PROMINENCE

CHROMOSPHERE

PHOTOSPHERE

CONVECTION ZONE

RADIATIVE ZONE

CORE

123 AU

HELIOSPHERE

SUN'S MAGNETIC BUBBLE

SUN'S COMPOSITION

H
≈70%

He
≈28%

≈2%
Other

Fig. 5

kinder, gentler, and more protective of humanity than we ever thought.

The sun's magnetic field, or *heliosphere,* is a magnetic bubble with a diameter of 246 au, which protects the whole solar system from the harmful effects of *cosmic rays*.

Our sun, like any other star, is made of a bubble of hot, dense plasma (ions and electrons) which makes it an almost-perfect sphere. The sun's huge mass creates a pressure equal to 3.8 trillion psi at its center, producing a temperature of 15 million °C (27 million °F), providing a suitable environment for the ignition of thermonuclear fusion. The creation of nuclear power makes it possible for any star such as our sun to stay stable, providing heat, light, and a magnetic field for it.

There are different layers in the sun's composition, starting with its *core* in the center, which contains 34% of the sun's mass, but only 0.8% of its volume. The next layer is the *radiative zone*, where the sun's energy is stored for millions of years before it is sent to the higher level called the *convection zone*.

From that point on, there are four more layers that are considered the sun's *atmospheres*. Starting from the bottom layer, they are the *photosphere,* the *chromosphere,* the *corona,* and the *heliosphere*.

The photosphere is where the sun's radiation appears as visible light, and it is also where *sunspots* originate. Sunspots are bundles of magnetic fields which, when unraveled, will release energy that goes through the chromosphere and the corona, appearing as *solar flares* (tongues of fire) and *prominences*. These will shoot up into space as far as half a

SOLAR SYSTEM'S PLANET'S MOONS & ORBITAL PERIODS

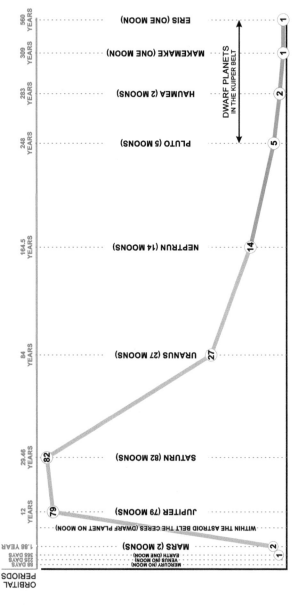

Fig. 6

million miles. When the solar flares and prominences cool down in the space, they become *solar wind*, which carries the sun's magnetic field all the way to the boundaries of the solar system, the *heliopause* (Fig. 5).

It is the solar wind that creates the sun's magnetic atmosphere's most outer layer, called the *heliosphere*. Since the chromosphere (sphere of color) and corona are not as bright as the photosphere, the only time that the chromosphere and corona are visible is during a *total solar eclipse,* when the chromosphere becomes visible as a *red rim* around the sun and the corona as a *halo* around the red rim.

THE PLANETS

Our solar system, with billions of moving pieces, is more precise, sophisticated, and complicated than a vintage Swiss watch. There are eight planets (Venus, Mercury, Earth, Mars, Jupiter, Saturn, Uranus, and Neptune), five dwarf planets (Pluto, Makemake, Ceres, Haumea, and Eris), and millions of asteroids of all different sizes and shapes. These asteroids move as a river of stones one AU wide and it is called the *asteroid belt,* located between Mars and Jupiter. Only one of the five dwarf planets, *Ceres,* is floating within the asteroid belt. There is also a river of frozen objects called the *Kuiper belt,* comprised of floating, *volatile,* frozen objects (methane, ammonia, water, and carbon dioxide). The Kuiper belt is twenty times wider than the asteroid belt and 200 times more massive. It is situated beyond Neptune's orbit.

Four of the five dwarf planets are within the Kuiper belt.

All these moving pieces have been harmoniously going around with a speed of thousands of miles per hour for the last 4.6 billion years.

Furthermore, simultaneously, the whole solar system has been orbiting around the Milky Way Galaxy's Sagittarius A* and its supermassive black hole. There are also 205 moons belonging to six planets, and nine moons orbiting the four dwarf planets (Fig. 6).

The speed of each planet around the sun is different, and it is called *differential rotation*.

The planets and the objects closer to the sun orbit faster than the objects farther from it. Therefore, they all have different annual rotations around the sun.

For an easier memorization of these planetary years, I have come up with a comparison of the human lifecycle with the planetary annual rotations around the sun. These numbers are very close, but not exact.

Mercury: 88 days (the first trimester)

Venus: 9 months (a whole pregnancy period)

Earth: 1 year

Mars: 2 and ½ years (a toddler)

Jupiter: 11 and ½ years (a sixth grader)

Saturn: 29 years (age of strength)

Uranus: 84.3 years (senior citizen)

Neptune: 166 years (2 x senior citizen)

The dwarf planet Pluto: 248 years (3 x senior citizen)

Furthermore, the whole solar system orbits around Sagittarius A* (the center of the Milky Way) every 225 to 250 million Earth-years, which is called a *cosmic year*.

FORMATION OF THE EARTH

After the solar nebula created the solar system, it took about another 100 to 200 million years before the planets got their mass, shape, and places in the solar system as they have today.

During this period, while the Earth was still a ball of fire, two distinct phenomena took place. *First*, the abundance of *hydrogen* and *helium* gases on the Earth's surface started to escape into the space around the Earth. However, after a period of time, the concentration of the hydrogen and helium atoms in space became dense enough for the Earth's gravity to hold them down, preventing them from escaping any further into the space; that became the first layer and the beginning of the formation of the Earth's atmosphere. *Second*, when the Earth's outer layer cooled down and hardened, it prevented the escape of gases and heat from underneath it.

Naturally, the trapped gases and heat had to escape, and the only way out was, and still is, through the phenomenon of *volcanic* eruptions. The volcanic lava from the deep was ejected to the Earth's surface, containing a variety of elements and gases, such as: water vapor (H_2O in a gas form), carbon dioxide (CO_2), and ammonia (NH_3).

When these elements reached the surface, with the help from the *sunlight*, simple bacteria and plants were produced. Then, in turn, these bacteria and plants went through the processes of *photosynthesis* using the CO_2 to produce oxygen.

(CO_2+H_2O+energy from sunlight=glucose+O_2)

Naturally, the volcanic activities were the free forces of nature, bringing the heavier elements, embedded deep

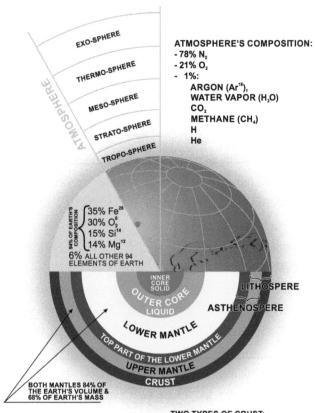

ATMOSPHERE'S COMPOSITION:
- 78% N_2
- 21% O_2
- 1%:
  ARGON (Ar^{18}),
  WATER VAPOR (H_2O)
  CO_2
  METHANE (CH_4)
  H
  He

EXO-SPHERE

THERMO-SPHERE

MESO-SPHERE

STRATO-SPHERE

TROPO-SPHERE

ATMOSPHERE

$\begin{cases} 35\% \ Fe^{26} \\ 30\% \ O_2^8 \\ 15\% \ Si^{14} \\ 14\% \ Mg^{12} \end{cases}$ 94% OF EARTH'S COMPOSITION

6% ALL OTHER 94 ELEMENTS OF EARTH

INNER CORE SOLID

OUTER CORE LIQUID

LOWER MANTLE

TOP PART OF THE LOWER MANTLE

UPPER MANTLE

CRUST

LITHOSPERE

ASTHENOSPERE

BOTH MANTLES 84% OF THE EARTH'S VOLUME & 68% OF EARTH'S MASS

TWO TYPES OF CRUST:
- CONTINENTAL CRUST - LIGHTER ELEMENTS
- OCEANIC CRUST - HEAVIER ELEMENTS

FIVE SPHERES OF EARTH:
1. ATMOSPHERE (AIR)
2. BIOSPHERE (LIFE)
3. HYDROSPHERE (WATER)
4. GEOSPHERE (CRUST, MANTLES, CORE)
5. MAGNETOSPHERE

GEOSPHERE:
- CRUST - LIGHT ELEMENTS SUCH AS SILICON, LESS THAN 1% OF THE EARTH'S MASS.
- MANTLES
- CORE - VERY DENSE ELEMENTS SUCH AS NICKEL & IRON

Fig. 7

in the planet, to the surface. The reason for the heavier elements sinking into the center of the planets is due to the fact that at the beginning of every planet's formation (500 million years into their formation), while it is still hot and all the elements are uniformly dispersed throughout the planet, a process called the *iron catastrophe* will take place that makes the *planetary differentiation* possible.

Planetary differentiation means that the heavier elements are deeper in the center of the planet while the lighter elements are at the surface.

There are *four forces* helping to make the *iron catastrophe processes* possible:

- *first,* the residual heat from the planet's formation
- *second,* the celestial body's own gravity (gravitational potential)
- *third,* the internal *frictions,* which cause heat
- *fourth,* the heat produced from the decay of *radioactive materials* inside the sphere itself.

These four forces create enough heat to facilitate the movement of the heavy elements, such as iron and nickel, from the top to the core of the planet, making the planetary differentiation possible.

Before we learn about the Earth's atmosphere, I must mention that there are five major, distinct spheres comprising the Earth as a whole.

THE EARTH'S FIVE MAJOR SPHERES

THE ATMOSPHERE (AIR)

We will talk about the Earth's atmosphere in detail later. However, in brief, it is good to know that the atmosphere starts a few feet under the ground and travels up to 1,000 km (620 miles) above sea level. It protects life on Earth from the harmful effects of the solar wind and CR. It also regulates the Earth's temperature. It is composed of 78% nitrogen and 21% oxygen, and the remaining 1% is H, He, CH_4, Kr, Ne, CO_2, and water vapor (H_2O).

THE HYDROSPHERE (WATER)

Water exists on Earth in three forms: liquid, solid, and gas. 71% of the Earth's surface is covered with water, of which 97% is oceanic salty water.

Most of the fresh water is frozen. Some scientists consider ice, glaciers, and icebergs in a category of their own and call it the *cryosphere*. Other scientists believe all three forms of water are part of the hydrosphere.

THE BIOSPHERE (LIVING THINGS)

This sphere includes microorganisms, plants, and animals living in different habitats, such as deserts, grasslands, and tropical forests.

Some scientists believe there is another sphere called the *anthroposphere*, or *technosphere*, that is not a part of the biosphere and should be considered as a separate sphere all by itself in the Earth's system. They assert that the anthroposphere / technosphere is everything that is manmade or modified by man. In 2016, scientists estimated that the total weight of the *human-generated structures* on Earth was 30 trillion tons, and they decided to name that zone the anthroposphere / technosphere.

GEOSPHERE (CRUST, MANTLE, CORE)

The geosphere is the rigid part of the Earth that includes the *crust*, which consists of the continental and oceanic crust making up less than 1% of the Earth's mass; the *mantle*, which is semi-fluid and rich in iron and nickel, making the movement of *tectonic plates* possible; and the *core*, which consists of the outer core and the inner core. The outer core is molten iron and nickel, while the inner core is made of solid iron and nickel.

There are two more *spheres* that are part of the geosphere: the *lithosphere* and the *asthenosphere*. Each specify a different depth of the Earth's solid shell. The lithosphere is a combination of Earth's crust and upper mantle, while the asthenosphere is a combination of the upper mantle and the top part of the lower mantle (Fig.7).

THE MAGNETOSPHERE (EARTH'S MAGNETIC FIELD)

After the *heliosphere* (the sun's magnetic field / bubble), the Earth's *magnetosphere* is the second source protecting the Earth's environment against the harmful ionizing radiation of the solar wind and cosmic rays.

The heliosphere is the sun's bubble-like magnetic field that protects the whole solar system by reflecting cosmic rays into space. The farthest point of the heliosphere from the sun is called the *heliopause*. Within the heliosphere, Earth's magnetosphere looks like a single scoop of ice cream on a cone (Fig. 8), lying on its side, with the top of the ice cream facing the sun and the cone pointing away from it. The magnetosphere is shaped by the sun's solar

EARTH'S MAGNETO-SPHERE

(not to scale)

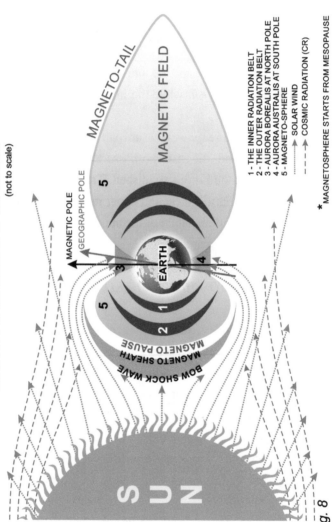

MAGNETO-TAIL

MAGNETIC FIELD

MAGNETIC POLE

GEOGRAPHIC POLE

EARTH

BOW SHOCK WAVE

MAGNETO SHEATH

MAGNETO PAUSE

SUN

1 - THE INNER RADIATION BELT
2 - THE OUTER RADIATION BELT
3 - AURORA BOREALIS AT NORTH POLE
4 - AURORA AUSTRALIS AT SOUTH POLE
5 - MAGNETO-SPHERE

→ SOLAR WIND
⇢ COSMIC RADIATION (CR)

* MAGNETOSPHERE STARTS FROM MESOPAUSE

Fig. 8

wind. The very top part of the ice cream that is facing the sun is called the *magnetopause,* and that is the point where the pressure of the solar wind and the pressure of the magnetosphere are in balance.

As the solar wind's pressure increases or decreases, the magnetopause moves inward or outward in response to it. On the opposite side of the magnetopause is the cone-shaped tail, called the *magnetotail*.

The magnetosphere starts from the *mesopause* (the highest point of the third layer of the atmosphere).

Furthermore, within the magnetosphere, there are two fat *radiation belts* around the Earth: the *inner radiation belt* and the *outer radiation belt*. The two belts are called the *Van Allen radiation belts*, discovered by Van Allen in 1958.

They are *depots* of trapped ionized particles from the solar wind and the cosmic rays that somehow managed to pass the magnetopause. The belt closest to Earth is the inner radiation belt, which contains all the trapped ionized *protons*, and the one farthest from the Earth is the outer radiation belt, which contains all the trapped ionized / energetic *electrons*.

Sometimes, a third radiation belt appears, but after a few weeks or months, it is blown away by the sun's shockwaves.

The phenomena of *auroras* are also related to the subject of the magnetosphere. At the North and South Poles where the magnetosphere is the thinnest and the most vulnerable to harmful radiation, the *solar storms* (coronal mass ejection) and the CR pass through the magnetosphere and collide with the atmospheric gases (oxygen or nitrogen) at the thermosphere level, creating a colorful cloud of ionized particles in the sky called *auroras*. If the collision is with the *oxygen gas*, it creates a display of red and green colors, and if the collision is with the *nitrogen gas*, then the clouds become an array of blues and purples. The aurora that takes place in the Arctic is called the Northern Lights, or *aurora borealis*, and the aurora that takes place in the Antarctic is called the Southern Lights, or *aurora australis*.

Any planet that has an atmosphere and a magnetic field can experience an aurora. In fact, there are pictures of auroras at the poles of the two planets of Jupiter and Saturn.

CHAPTER THREE

"We have forgotten how to be good guests,
how to walk lightly on the earth
as its other creatures do."

—Barbara Ward

THE EARTH'S ATMOSPHERE

The Earth's atmosphere is the *third* and *last* protective shield for the Earth against the ionizing radiation of the solar wind and the CR. The other two being the heliosphere—the sun's magnetic field—and the magnetosphere—the Earth's magnetic field.

The *Earth's atmosphere* is composed of five layers and is a living, breathing, expanding, contracting, sensitive, nurturing, protective piece of heaven who spreads her blessed wings upon our beloved Earth. It is the ultimate blanket that took Mother Nature billions of years to perfect. There are some other atmospheres elsewhere in the visible universe; however, none of them are as sophisticated and delicate as our atmosphere. It is a delicate system, and it should be treated as such.

Unfortunately, since the Industrial Revolution of the 1750s, we've acted like a mad bull in a china shop, destroying the very air we need to breathe, the water we need to drink, and the land we need to cultivate.

We are destroying our ecosystem like there is no tomorrow and there are no next generations. What kind of intelligent beings are we? Who do we think we are to give ourselves the right to destroy the things that we should be safeguarding for many generations to come? Humanity must start thinking before it is too late!

Remember when the coronavirus arrived? A tiny virus that threw a wrench into the unstoppable, huffing and puffing machinery of the world, bringing the whole world to a halt! We must learn from every disaster that we are weaker and more fragile than we think we are, and that the wrath of nature will be frightening and costly.

The Earth's atmosphere is one of the nature's most crucial elements that makes life possible here. Just a glance at Fig. 9 can tell us a good deal about the Earth's atmosphere. It is made of five distinct layers, each with a specific thickness, temperature, and function.

The five layers of the atmosphere are divided into two major categories:

First, the *neutrosphere*, which contains the two levels closest to Earth (troposphere and stratosphere), extends up to 50 km and is electrically neutral.

Second, the *ionosphere,* which contains the three upper

EARTH ATMOSPHERE

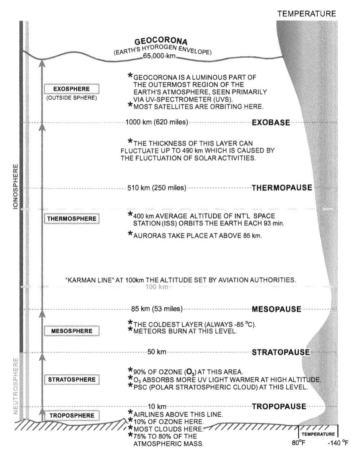

TEMPERATURE

GEOCORONA
(EARTH'S HYDROGEN ENVELOPE)
65,000·km

EXOSPHERE
(OUTSIDE SPHERE)

*GEOCORONA IS A LUMINOUS PART OF THE OUTERMOST REGION OF THE EARTH'S ATMOSPHERE, SEEN PRIMARILY VIA UV-SPECTROMETER (UVS).
*MOST SATELLITES ARE ORBITING HERE.

1000 km (620 miles) **EXOBASE**

*THE THICKNESS OF THIS LAYER CAN FLUCTUATE UP TO 490 km WHICH IS CAUSED BY THE FLUCTUATION OF SOLAR ACTIVITIES.

510 km (250 miles) **THERMOPAUSE**

THERMOSPHERE

*400 km AVERAGE ALTITUDE OF INT'L SPACE STATION (ISS) ORBITS THE EARTH EACH 93 min.

*AURORAS TAKE PLACE AT ABOVE 85 km.

"KARMAN LINE" AT 100km THE ALTITUDE SET BY AVIATION AUTHORITIES.
100 km

85 km (53 miles) **MESOPAUSE**

MESOSPHERE

*THE COLDEST LAYER (ALWAYS -85 °C).
*METEORS BURN AT THIS LEVEL.

50 km **STRATOPAUSE**

STRATOSPHERE

*90% OF OZONE (O_3) AT THIS AREA.
*O_3 ABSORBS MORE UV LIGHT WARMER AT HIGH ALTITUDE.
*PSC (POLAR STRATOSPHERIC CLOUD) AT THIS LEVEL.

10 km **TROPOPAUSE**

TROPOSPHERE

*AIRLINES ABOVE THIS LINE.
*10% OF OZONE HERE.
*MOST CLOUDS HERE.
*75% TO 80% OF THE ATMOSPHERIC MASS.

IONOSPHERE

NEUTROSPHERE

TEMPERATURE
80°F   -140 °F

Fig. 9

layers of the atmosphere (mesosphere, thermosphere, and exosphere) and extends from the altitude of 50 km (31 miles) up to 65,000 km (40,389 miles) into space.

The ionosphere plays an important role in atmospheric electricity and radio wave propagation to distant places on Earth.

THE FIVE LAYERS OF THE EARTH'S ATMOSPHERE

FIRST LAYER: THE TROPOSPHERE

It starts a few feet underground and extends up to 10 km altitude, to the point that is called *the tropopause*.

It contains 10% of the atmosphere's *ozone* (O_3), while the other 90% of ozone is in the upper part of the next layer, the *stratosphere*.

The troposphere is the narrowest and densest layer of the atmosphere, containing 75–80% of the atmosphere's mass.

All weather activities take place in this layer. It is the wettest layer, containing 99% of the *water vapor*. Water vapor is water in its invisible gaseous form, and when it is condensed at the right temperature and pressure, it will appear as fog. The high concentration of the water vapor in the air is called *humidity*. Temperature can affect the troposphere's thickness around the globe; it is twelve miles thick around the equator and four miles thick at the poles, swelling in heat and contracting in cold weather The temperature of the troposphere will drop at the higher altitudes. The border between the troposphere and the stratosphere is called the tropopause. As a matter of fact, temperature is one of the main factors for the distinction between the borders of each of the two adjacent layers of the atmosphere.

Water vapor also accounts for 97% of total greenhouse warming.

SECOND LAYER: THE STRATOSPHERE

The stratosphere is the second layer of the atmosphere from the ground, starting from 10 km and extending up to 50 km altitude. Unlike the troposphere that gets colder at a higher altitude, in the stratosphere, the higher the altitude, the hotter it gets, because 90% of the atmosphere's ozone (O_3), is located at the 2/3 upper level of the stratosphere (the other 10% of the ozone is in the troposphere). The ozone traps the *heat and energy* from the UV light coming from the sun, making the upper portion of the stratosphere hotter. The ozone is a blessing in the upper level of stratosphere (trapping the bad UV radiations), while it is a nuisance in the troposphere, causing *eye and lung irritation*.

In the upper portion of the stratosphere, at the sites of the two polar zones where there is a concentration of ozone, a phenomenon can take place that can cause the depletion of the *ozone layer* (known as ozone hole).

The phenomenon is called *polar stratospheric clouds* (PSCs).

There are two types of PSCs: the *harmless* one and the *harmful* one. For a PSC to form, three favorable conditions must exist.

First, it must take place at two locations of north or south poles, where the magnetosphere is the thinnest and most vulnerable to the harmful radiation.

Second, it must take place during the winter when the temperature is extremely cold (-78 °C).

Third, the water droplets must be available.

In these types of conditions, the water droplets will freeze into ice crystals which are called *nacreous* or *mother of pearl*. This type of PSC is harmless to the ozone layer. However, if *nitric acid* (HNO_3) is present in said favorable conditions, then the reactions between the PSC's particles will cause the formation of a highly reactive *chlorine gas* (Cl), which is a destroyer of the ozone (O_3).

Finally, in severe cases, when the ozone layer is depleted, it is called the *ozone hole*.

THIRD LAYER: THE MESOSPHERE

This layer starts from the *stratopause* at 50 km from the sea level and extends up to 85 km altitude. One of the main elements that distinguishes the mesosphere from its lower and upper layers is the temperature. Unlike the stratosphere, which gains heat as it goes to the higher altitude, the mesosphere steadily loses heat as it rises up to its mesopause.

One of the most important functions of the mesosphere is to burn and destroy *meteors*, due to the friction of its high-density gases.

FOURTH LAYER: THE THERMOSPHERE

The air density in this level is so low that some scientists call it *outer space*. The thermosphere starts from 85 km (53 miles) and the highest it can go is 1,000 km. Within this layer, from the altitude of 160 km (99 miles) and up, the molecular interactions are so infrequent that the transmission of soundwaves is not possible due to lack of mediums such as air, liquids, or solids. Also in this layer, the higher the altitude, the hotter it gets because of intense solar activities and ionization processes. In some areas, the temperature will rise to 4,500 °F. One piece of good news from this layer is that the charged (ionized) particles can refract *radio waves* to be received beyond the horizon.

This is the level that the space shuttle and the International Space Station (ISS) are flying (at a 400-km or 250-mi-average altitude). It is also the layer where the colorful *aurora borealis* will mesmerize the Inuit people over the North Pole, and the *aurora australis* will dance over the South Pole, entertaining the penguins.

The layer of the thermosphere starts at the mesopause (85 km altitude) and extends to either 510 km or 1,000 km high. There is a 490-km discrepancy between these two numbers; it is because the *thermopause* will inflate and deflate within this 490 km depending on the sun's activities. The thermosphere's temperature increases with the altitude due to the absorption of more highly energetic solar radiation. Also in this layer, the *heat* and *expansion* of the thermopause region causes a *drag* on satellites, causing them to fall from the sky; the engineers must boost them up in order to keep them in orbit.

There is one more important point concerning the thermosphere and that is the *Kármán line*. It is a line at 100 km above sea level, only for the purpose of aeronautic regulations.

FIFTH LAYER OF THE ATMOSPHERE: THE EXOSPHERE

The fifth and last layer of the Earth's atmosphere is the exosphere (outside sphere). It is composed of the lightest gases, such as hydrogen and carbon dioxide. These gases are still gravitationally bound to Earth; however, they are so dispersed that there is no collision among them.

The exosphere starts from thermopause (exobase) and covers up to an altitude of 65,000 km or 40,000 miles. The exosphere has no *exopause* region, due to its lack of measurable air molecules. However, at the outer boundary of the exosphere, there is a *hydrogen envelope* between our atmosphere and the sun, which is called the *geocorona*. It extends for almost 60 miles into space, and it is only seen by an ultraviolet spectrometer (UVS).

EXCESS GREENHOUSE GASES, GLOBAL WARMING, AND THE OZONE HOLE

These three issues affect the Earth's atmosphere.

Greenhouse gases (GHG) are the type of gases that absorb the sun's heat and radiation in the atmosphere to keep the atmosphere and the Earth warm. However, since the Industrial Revolution of the 1750s, the production of GHGs has increased. As a result, the excess amount of GHGs in the atmosphere contributed to a higher absorption and retention of UV light coming from the sun and CRs, causing the atmosphere to become warmer and creating the phenomenon of *global warming*.

The Industrial Revolution created a new lifestyle for us—the running factories, burning fossil fuels, cutting down the forests, new agricultural practices, mining, drilling oil, livestock, and decay of organic wastes—all of which led to the increased emission of GHGs. There are several GHGs, including carbon dioxide (CO_2), nitrous oxide (N_2O, also known as laughing gas or oxide of nitrogen), methane (CH_4), and halogenated gases that are highly reactive, such as chlorine, fluorine, bromine, and iodine.

There are two other main GHGs: ozone (O_3), and water vapor ($H2O$ in an invisible gaseous form). However, since the ozone is so instrumental in absorbing the UV lights from the sun, it is dismissed as a GHG. But the water vapor is a different story. Although only 1% of the atmospheric gases constitute the GHGs (the other two major constituents gases of the atmosphere are 78% nitrogen and 21% oxygen), the water vapor's contribution to global warming is the most effective one. The process is very simple. Water on Earth absorbs the heat and becomes water vapor, and, as a gas, rises into the atmosphere. Then water vapor, in the colder and higher altitudes of the atmosphere, will condense to become a cloud, mist, or fog, and then rain, releasing the heat back into the atmosphere, and that contributes to the warming of our atmosphere and Earth; in extreme cases, the abundance of GHGs will cause *global warming*.

Furthermore, as mentioned before, there is a phenomenon called *polar stratospheric clouds* that takes place in the north and the south poles of the Earth in the stratosphere. This phenomenon is not harmful to the environment; however, if nitric acid (HNO_3)—which is a highly corrosive mineral acid—is added to this combination, then it will produce chlorine gas (Cl). Chlorine gas is an ozone destroyer, and in severe cases, a complete depletion of ozone is termed as a *hole* in the ozone layer.

There is not an actual hole, but rather a lack of O_3 concentration in that part of the space to absorb harmful UV light.

Understanding the disastrous environmental impacts to our atmosphere and the global warming caused by GHGs has alerted the world community to come up with some guidelines, such as international mandates by the *Montreal Protocol*, and, in the United States, the *Clean Air Act*, meant to cut down on industrial gases that contribute to extreme production of the GHGs!

CHAPTER FOUR

"Light is the ultimate messenger of the universe."
—BBC World Service

ELECTROMAGNETIC RADIATION SPECTRUM

Earlier, we read about *cosmic radiation / rays (CR)* from outer space and the solar wind from the sun heading toward the Earth (only 5% of the sun's output reaches Earth). The Earth's atmosphere will filter out most of the *harmful ionizing radiations* and let in those that are *beneficial radiation* to life.

How do we categorize and differentiate between different radiation with different energies and characteristics? That's where the *EM radiation spectrum* comes to the rescue. Through a technique called *spectroscopy*, different categories of radiations are separated only on the basis of different frequencies. These frequencies range from one cycle per second up to octillion (10^{27}) cycles per second.

But before we get into the categories of EM radiation,

EM RADIATION SPECTRUM

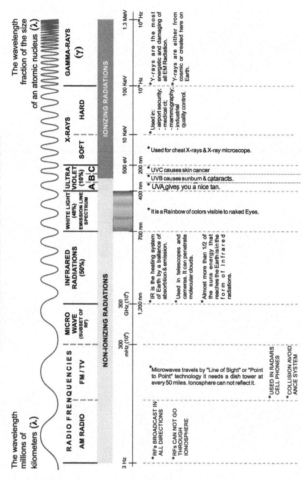

Fig. 10

we must define *EM radiation*, *frequency*, and *wavelength*.

First: EM radiation is made up of self-propagating waves of two synchronized, oscillating perpendicular fields of *electric* and *magnetic* fields. The fields move in a third direction in space-time, with the speed of light (c), in a vacuum.

Second: frequency (f) is the number of cycles per second, and the standard unit of frequency is called Hertz (Hz).

Third: wavelength (lambda / λ) is the distance between the two successive crests or troughs of a sine wave.

Wavelength and frequencies are inversely related to one another.

As the *number* of the frequencies (f/v) goes up in the EM spectrum, the *size* of the wavelength goes down. At the beginning of the spectrum, the size of the wavelength will start from hundreds of miles per second to a few nanometers per second at the end.

The table of the EM radiation spectrum is divided into seven distinct *bands* of frequencies.

The sole criterion for the divisions between these categories is their frequencies. The higher the frequency of the radiation, the higher the energy, and higher energy means stronger ionizing power.

THE SEVEN BANDS OF EM RADIATION

EM radiation is divided into seven bands of frequencies:
1- Radio frequencies (RFs)
2- Microwave frequencies
3- Infrared frequencies
4- Visible light (white light)

5-   Ultraviolet frequencies

6-   X-rays

7-   Gamma rays

Each of these bands of frequencies has different characteristics that determine how they are produced or will interact with matter, and their practical applications in different industries.

In terms of their *strength* and *practicality,* all seven bands of radiation are divided into two main groups of *non-ionizing* and *ionizing* radiation.

Non-ionizing radiation is the type of radiation that doesn't possess enough energy to knock an electron out of another atom's orbit to make an ion (the atom that loses or gains an electron becomes ionized). On the other hand, ionizing radiation is the type of radiation that possesses enough energy to overcome the *binding energy* of the electron(s), knocking it out of its orbit and turning that atom into an *ion*.

On the scale of the EM spectrum, non-ionizing radiation starts from the RFs and extends to microwave frequencies, infrared frequencies, visible light, and the first, lower half of the ultraviolet band.

The *ionizing* radiation starts from the second, upper half of the ultraviolet band and extends to X-rays and gamma rays. Ionizing radiation can ionize other atoms or damage the human's cell structures (by damaging the DNA).

RADIO FREQUENCIES (RF)

These frequencies range from 3 Hz up to 300 MHz, which includes AM/FM radio and TV channels. RFs

EM RADIATION SPECTRUM

OUR FIVE SENSES

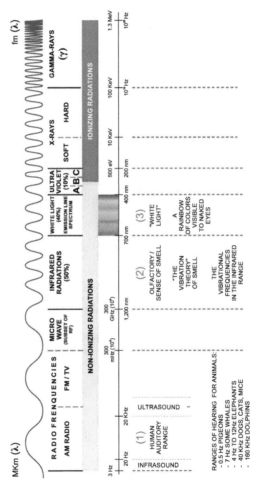

Fig. 11

broadcast in all directions (as opposed to microwave frequencies that travel only by *line of sight* or *point to point* by a narrow beam).

MICROWAVE FREQUENCIES

Microwave frequencies range from 300 MHz up to 300 GHz, and they are used in microwave ovens, radars, cell phones, satellites, space craft communications, collision-avoiding systems, garage openers, and many more remote-controlled devices. But the only shortcoming about microwave frequencies is that they only travel by line of sight or point to point.

Note: before it gets a bit confusing, I must explain that the scientists would like to deal with different categories of EM radiation in different *units* of measurements.

For example, they use the unit of *frequency* (Hz) for RFs and microwaves, while they prefer *nanometer wavelength (λ)* for infrared, visible light, and ultraviolet radiations. In the categories of X-rays and gamma rays, they prefer the units of *electron volts* (eV).

INFRARED FREQUENCIES

If the width of all three bands of infrared, white light, and ultraviolet were collectively 100%, the width of infrared would be 50%, white light would be 40%, and the remaining 10% would go to the ultraviolet band. That shows how important the infrared band of radiation is in our daily lives. Infrared radiation contains the most important frequencies for keeping the atmosphere, oceans, and Earth warm. The infrared radiation band's wavelength extends from 1,200 nanometers (nm) to 700 nm. (Notice that the higher the frequencies (f/v) of the infrared radiation go, the shorter the wavelength becomes.)

Infrared frequencies are also used in astronomical research here on Earth to observe the night's activities with infrared cameras.

VISIBLE LIGHT (WHITE LIGHT)

The band of white light ranges from 700 nm wavelength (λ) to 400 nm. In this band of frequencies, the seven main colors of the rainbow and a thousand different shades of colors in between are visible to our naked eyes. The seven colors range from the lower frequencies (longer wavelength) to the higher frequencies (shorter wavelength). The visible colors include red, orange, yellow, green, blue, indigo, and violet.

ULTRAVIOLET FREQUENCIES

Ultraviolet light is not visible to the naked eye because its wavelength is shorter than the wavelength of visible light. The UV band ranges from 400 nm wavelength down to 200 nm.

The UV band is divided into three sub-bands of UVA, UVB, and UVC.

UVA is not damaging; it is instrumental in the skin's formation of Vitamin D. It also provides the sun-lovers with a nice tan. UVA is the last band of frequencies that our atmosphere will allow to reach Earth, because it is not ionizing.

UVB is the beginning of the ionizing radiation. Most of it is absorbed by the atmosphere's ozone layer (at the upper level of the stratosphere). Unfortunately, some of it will pass through the atmosphere causing *cataracts* of the eye and *sunburns* of our skin.

UVC is the most damaging UV radiation, causing skin cancers and other health problems. However, if our ozone layer in the upper level of the stratosphere is still intact, it will absorb most of it before it reaches Earth.

X-RAYS

As I mentioned earlier, from this point on, the scientists would rather deal with the ionizing radiation in terms of their strength in electron volts (eV), rather than dealing with them in terms of frequencies (Hz) or nanometer wavelength (λ).

Electron volt (eV) is the unit of energy. One eV is the amount of *kinetic energy* that one electron will gain from rest position in a vacuum, between the cathode and the anode with a potential energy of *one volt.*

One eV=1.6×10^{-19} J (joule)

The X-ray band is divided into two sub-bands of *soft X-rays,* ranging from 500 eV to 10 KeV, and *hard X-rays,* ranging from 10 KeV to 100 KeV. The soft X-rays are being used in hospitals for imaging of all types of internal organs and are also being used in X-ray microscopes.

The hard X-rays are in higher-frequency brackets, possessing more penetrating power or electron volts (eV). Hard X-rays are used in hospitals for mammography and computed tomography (CT). They are also used in industrial quality controls. And finally, the most powerful X-rays are used by airport security.

GAMMA RAYS

Gamma rays are the most powerful form of ionizing EM radiations that are being used in hospitals and different industries.

There are two sources of gamma radiation:

First, there is the gamma radiation coming from the sun and the CR. They are the most powerful gamma rays produced by the *hottest* and the *most massive stars*, such as neutrons, pulsars, quasars, and black holes. Furthermore, during solar flares or solar prominences, we get a burst of gamma radiation that is closely associated with the ejection of plasma and charged particles through the sun's corona into outer space. If a solar flare or solar prominence is in the direction of the Earth, the charged particles can penetrate through the *ionosphere* (the three layers of mesosphere, thermosphere, and exosphere), reaching the thermosphere and displaying auroras at the North and South Poles; this can also sometimes disrupt long-range radio communications.

Second, the gamma sources on Earth, which are:

A. *Lightning.* Lightning is the buildup of electrons and positrons inside the clouds (static electricity), followed by an annihilation of the two, which produces gamma radiation.

B. *The radioactive decays.* When radioisotopes decay, the energy emitted from their nuclei is in a form of gamma radiation.

C. *Nuclear power plants.* Nuclear power plants produce

gamma radiations during the processes of producing electricity. These nuclear power plants are environmentally a favorable source of clean energy unless there is an accident.

Unfortunately, accidents happen, and have caused death and destruction.

There have been a few disastrous cases, such as the Three Mile Island accident in the US in 1979, and in 1986, in Russia, the *Chernobyl disaster,* that took place, taking an estimated 4,000 lives.

In 2011, in Japan, at *Fukushima Daiichi Nuclear Power Plant*, a 9.0-Richter-magnitude earthquake created a 15-meter-high (almost 50 feet) tsunami that disabled the power supply running the reactor's cooling system, causing panic and destruction.

Note:

These seven bands of EM radiation are not 100% separated, and there are some overlaps of frequencies by the adjacent bands.

A HUMAN'S FIVE SENSES UNDER THE UMBRELLA OF EM RADIATION

It is interesting and important to be able to pinpoint exactly where our three senses of *hearing*, *smell*, and *vision* are located within the EM spectrum. (The other two senses of *taste* and *touch* are said to be mechanical.)

I would like to call our three senses of hearing, smell, and vision our *wireless senses,* although they receive their frequencies through different ways of transmission. Vision is possible through EM radiations; hearing is possible through transmission of longitudinal soundwaves through air, liquid, or solid molecules; and the sense of smell is by the passage of microscopic molecules of smell through the nasal cavity.

THE SENSE OF HEARING

Humans can hear frequencies between 20 Hz and 20 KHz—the audible range.

Any frequencies below the audible range are called *infrasound,* which humans cannot hear, but some animals, such as pigeons, can hear as low as 0.5 Hz; some whales can hear as low as 7 Hz; and an elephant's hearing ranges from 4 Hz up to 12,000 Hz.

Above the human auditory range is *ultrasound.* These frequencies are also not audible to human beings, while mice, cats, and dogs are able to hear sounds up to 40,000 Hz; dolphins and bats will hear frequencies as high as 160,000 Hz.

How are audible frequencies heard? And what is the definition of sound?

Soundwaves (pressure waves) are the vibrations caused by the movements of any medium's molecules. There are different types of mediums, such as air, liquids, or solids. The denser the medium, the faster the soundwaves will travel (i.e. sound travels faster through diamond than air).

What makes hearing possible is *resonance.* Resonance occurs when the wavelength of a source of a sound is an integral fraction of the wavelength of the receiver of the sound. For example, if we create a continuous soundwave of one specific frequency and build a set of strings with different lengths under tension (receivers), the string that has the same resonance wavelength will *resonate* and its *vibration* will be visible.

The way we can hear different soundwave frequencies (wavelengths) is as follows: when the *soundwaves* reach the eardrums of both ears, it will cause the vibrations of eardrums with the same frequencies carried by the air molecules. These vibrations will be transferred to the oval window at the beginning of the cochlea, by the three bones in the middle ears.

The *cochlea* is a tapered channel filled with fluid, wider at the beginning and narrowed at the end of the cochlea channel. The cochlea is coiled into a spiral shape for compactness. When the stimulating soundwave reaches the oval window, it will be transmitted into the cochlea channel. The higher frequencies (pitch) are able to move the *hair cells* at the wide section of the cochlea channel, while the lower frequency soundwaves are able to travel deeper into the cochlea channel to move the hair cells at the narrower section.

Note: the frequency or pitch of a soundwave determines the movement of the hair cells at a different location(s) in the cochlea channel, while the amplitude or loudness of soundwave involves the number of hair cells being moved. The louder the sound, the higher number of hair cells will be resonated or moved.

And finally, the *physical movement* of the *hair cells* in the cochlea channel wall will stimulate the auditory nerves, which will convert the physical movement into an electrical signal. The auditory nerve will take the information into the brainstem and hypothalamus, and finally to the *cerebral cortex of the brain* for the recognition of different sounds.

THE SENSE OF SMELL

There are a few complicated theories in explaining how our sense of smell works.

However, the theory that I am interested in is the *vibration theory of olfaction*. It says that every *molecule* in nature possesses a specific frequency called *characteristic frequency*. This specific frequency is due to specific vibrations / frequencies amongst the constituent atoms in a molecule. These frequencies are in the range of infrared.

The theory posits that when microscopic molecules of smells are suspended in the air and we inhale them through our nose, the molecules will go through the dark smelling chamber of our nasal cavity.

There, at the roof of the nasal cavity, in an area the size of a stamp, sits the olfactory epithelium with its olfactory neurons and odor receptors. Each odor receptor is sensitive to a specific frequency within the infrared frequencies.

At this point, when the microscopic smell molecules contact the right odor receptors, the phenomenon of resonance / synchronization will take place and the signals will be sent to the brain for smell recognition.

Some animals have a better sense of smell because they have more olfactory receptors and longer smelling chambers.

THE SENSE OF SIGHT (THE COLOR VISION)

The most important sense in our body is the sense of sight—seeing the outside world through our eyes. With just a glimpse of scenery, our eyes receive thousands of different EM frequencies, each carrying a specific message.

The human eye can only see a small band of frequencies within the EM spectrum with the wavelength ranging from 700 nm to 400 nm. It is called the *visible* or *white light*.

Different colors and shades of colors are only separated from one another by a fraction of frequencies or wavelength.

For example, all shades of the color *red* range within 700s nm; *orange,* within 650s nm; *yellow*, 600s nm; *green,* 550s nm; *blue,* 500s nm; *indigo,* 450s nm; and finally, *purple,* 400s nm.

A combination of all these frequencies gives us white light (a full rainbow of colors*).* Any wavelength longer (lower frequencies) than white light is within the infrared band, which we cannot see (we can only feel these frequencies as heat); any wavelength shorter (higher frequencies) than white light is within the ultraviolet radiation band, which we also cannot see.

Now, how do we see things? How do our eyes convert different frequencies into a beautiful, panoramic image?

Let's do something crazy! Let's pack up a very light backpack and fly to Mumbai, India!

Imagine we are standing on the balcony of a five-star hotel in Mumbai, sipping our *bhang* and watching the Holi festival down below, where thousands of people are having fun on a super-crowded street. It is a festival of colors, love, and spring celebration. All types of vivid, bright, and beautiful colors are displayed. People are covered with splashes of different, vibrant colors. It is a sight to be seen.

Now, let's see how our eyes process this beautiful tapestry of colors.

All of what we see before us is nothing but different frequencies of EM radiation. A barrage of colors rushes into our eyes, going through our eye's *cornea*, then *pupil*, then through the *lens*, where it is bent and focused upon the *retina*. The *retina* is where a group of *light-sensing* cells called *photoreceptors* is located. There are two types of photoreceptors: *rods* and *cones*.

Rods are sensitive to dim lights, while cones are better converters of bright lights.

There are three types of *cone receptors*: red, green, and blue cones.

Each cone is sensitive to white light as well as being *more* sensitive to its own color frequencies.

Just like a color TV, which uses only three colors—red, green, and blue—to make thousands of shades of different colors, these *three different color cones* are also able to mix and match all the various incoming signals and frequencies (colors) to come up with the exact image that is out there. Just like a master painter who mixes different colors on his palette to get the exact colors of a beautiful butterfly he wants to paint, the color cones do just that by being most sensitive to their own colors and less sensitive to other colors. When a color image is focused on these cones, different types of cones will be *activated*. For example, if there is only a blue color, the blue cone receptors will be involved 100%; if there is only a red color, the red cone receptors will be activated 100%. For an image of mixed colors with more red and less green, the red cone receptors will be involved 80% and the green cone receptors will be involved 20% (approximations), and so forth. So, those three types of cones plus rods will mix and blend colors, either individually or by a different percentage of activation, to come up with an exact duplicate of a colorful image.

After the recognition of each specific frequency by the rods and cones located at the retina, the *optic nerves* will carry the information to the *visual cortex* in the back of our brain.

And voila! There it is! The most beautiful display of colors in a Holi festival!

CHAPTER FIVE

"Time goes, you say?
Ah no!
Alas, time stays, we go."
—Henry Austin Dobson

THE PERIODIC TABLE OF ELEMENTS

In 1869, Russian chemist and inventor *Dmitri Mendeleev* formulated *the periodic table of elements*. However, at that time, his table of elements contained only 10 elements (now we have 118 elements), with a prediction of four other elements based on their *atomic number* (the number of protons in their nuclei), *melting point,* and *density* (density is mass divided by volume: d=m/v).

Among his predicted elements, gallium (Ga-31) was discovered six years later, and then the next group of elements—scandium (Sc-21), germanium (Ge-32), and technetium (Tc-43)—were discovered thirty years after his death.

The periodic table of elements we have today is the

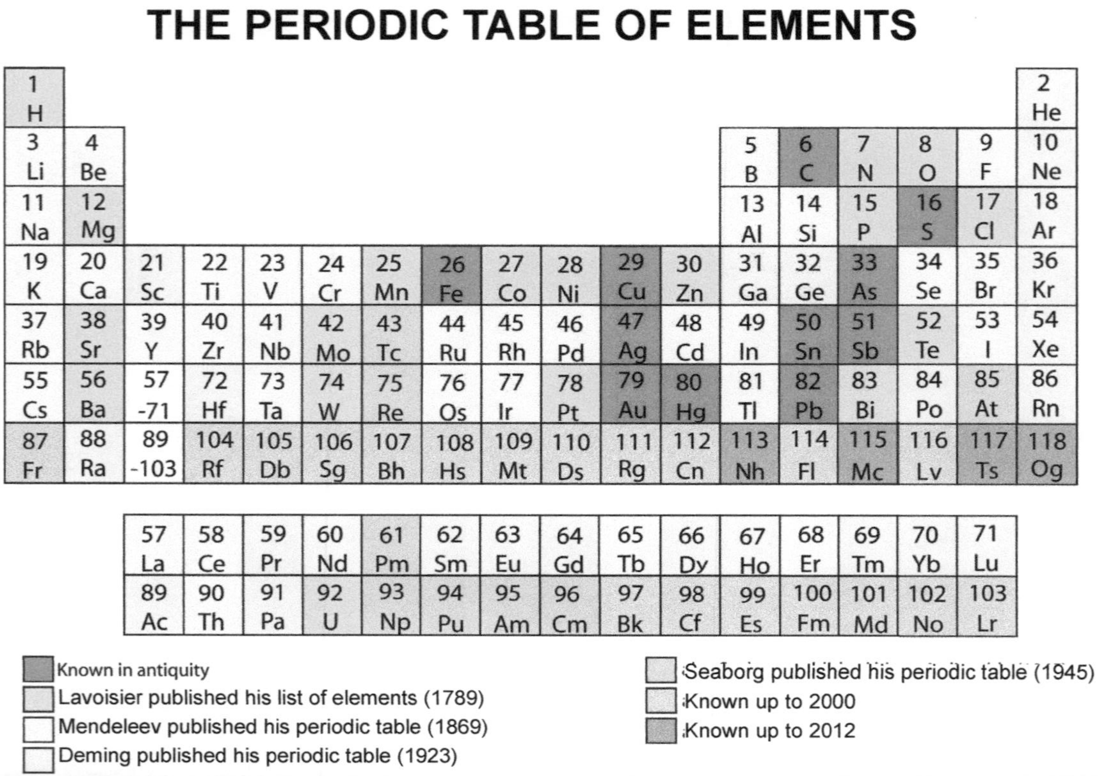

THE PERIODIC TABLE OF ELEMENTS

1 H																	2 He
3 Li	4 Be											5 B	6 C	7 N	8 O	9 F	10 Ne
11 Na	12 Mg											13 Al	14 Si	15 P	16 S	17 Cl	18 Ar
19 K	20 Ca	21 Sc	22 Ti	23 V	24 Cr	25 Mn	26 Fe	27 Co	28 Ni	29 Cu	30 Zn	31 Ga	32 Ge	33 As	34 Se	35 Br	36 Kr
37 Rb	38 Sr	39 Y	40 Zr	41 Nb	42 Mo	43 Tc	44 Ru	45 Rh	46 Pd	47 Ag	48 Cd	49 In	50 Sn	51 Sb	52 Te	53 I	54 Xe
55 Cs	56 Ba	57 -71	72 Hf	73 Ta	74 W	75 Re	76 Os	77 Ir	78 Pt	79 Au	80 Hg	81 Tl	82 Pb	83 Bi	84 Po	85 At	86 Rn
87 Fr	88 Ra	89 -103	104 Rf	105 Db	106 Sg	107 Bh	108 Hs	109 Mt	110 Ds	111 Rg	112 Cn	113 Nh	114 Fl	115 Mc	116 Lv	117 Ts	118 Og

57 La	58 Ce	59 Pr	60 Nd	61 Pm	62 Sm	63 Eu	64 Gd	65 Tb	66 Dy	67 Ho	68 Er	69 Tm	70 Yb	71 Lu
89 Ac	90 Th	91 Pa	92 U	93 Np	94 Pu	95 Am	96 Cm	97 Bk	98 Cf	99 Es	100 Fm	101 Md	102 No	103 Lr

Known in antiquity
Lavoisier published his list of elements (1789)
Mendeleev published his periodic table (1869)
Deming published his periodic table (1923)

Seaborg published his periodic table (1945)
Known up to 2000
Known up to 2012

Fig. 12

work of more than 133 years (1869 to 2002) of research and experiments by chemists and nuclear scientists. The last element in the table is the element oganesson (Og-118), discovered in 2002 by the Russian-Armenian nuclear physicist Yuri Oganessian.

WHAT IS AN ELEMENT?

An *element* is a substance that is made of one type of atom.

In the periodic table of elements, only the first 92 elements (up to element U-92) exist naturally on Earth; the rest are synthetically made in laboratories, cyclotrons, and nuclear reactors around the world.

Of all the 92 naturally existing elements on Earth, 94% are composed of *four* elements: 35% *Fe*, 30% O_2, 15% *Si*, and 14% *Mg*.

The remaining 6% is composed of different percentages of all the other elements.

Just by looking at the periodic table of elements, you are looking at the tip of an iceberg. Each element you see has one to tens of *isotopes*—and there are around 3,800 different types of isotopes.

WHAT IS AN ISOTOPE?

Each element has a few isotopes, and all behave the same because they all have the same atomic number (same number of protons in their nuclei) and the same number of electrons (therefore, the same *chemical properties*). The only difference is their number of neutrons, which will change the *atomic mass number* (n+p=a).

Isotopes of the same element have the same chemical properties because their number of electrons are the same; however, different isotopes have different half-lives (Tp) and different strengths (eV). For example, the element iodine has 37 isotopes, but only I-127 is stable and the rest are radioactive. All 37 of them can be utilized by the thyroid gland in the same way, but different isotopes of iodine are used for different purposes because they all have different half-lives and different strengths.

There are about 3,800 different types of isotopes; however, only 200 of them are being used by different industries on a daily basis.

To understand the properties of an isotope in more detail, one must understand the structure of an atom as well.

HISTORY OF THE ATOM

The *atom* has a long theoretical and experimental background.

Its journey started around 450 B.C., when a Greek philosopher named *Democritus* talked about the concept of the atom by simply stating that if you cut an apple into small pieces, at one point, you will come up with a piece of apple that is "uncuttable." And he called that piece "atom." However, the academia of the time found the idea too simplistic, and the idea was put to rest for almost 2,250 years.

Then, around the year 1800, a British chemist named *John Dalton* reviewed Democritus' idea of an atom, and through research, he came up with his own version of atomic theory. It states:

1. All substances are made of atoms.
2. Atoms are the smallest particles of matter.
3. They cannot be divided any further.
4. They cannot be created nor destroyed.
5. All atoms of the same element are the same.
6. Atoms join to form compounds.

Dalton's idea was soon widely accepted, and some of his points are still valid.

In 1865, *James Clerk Maxwell,* a Scottish scientist who is considered to be the third-greatest physicist after Newton and Einstein, formulated the theory of EM radiation, stating that *electricity, magnetism*, and *light* are different manifestations of the same phenomenon. Maxwell's discoveries led to the foundation of radio frequencies, electrical engineering, Albert Einstein's special theory of relativity, and Niels Bohr's quantum mechanics theory. Maxwell also demonstrated that EM radiation travels as a *wave* with the speed of light (c) in a vacuum.

In 1897, a British physicist named *J.J. Thomson* discovered *electrons*. Thomson posited that since electrons have a negative charge and atoms are neutral, then there must be a positively charged particle in an atom that makes the atom neutral. Therefore, he believed electrons were floating in positively charged, jellylike spheres. This theory is known as *the plum pudding model*.

In 1900, *Max Planck,* a German physicist, experimentally showed the effects of radiation on a *black body substance* (an opaque non-reflective body) on the atomic level, which exhibited the *characteristics* of matter. That showed Planck that, unlike the old belief by the classical physicists that energy is a continuous *wavelike* phenomenon (theorized by James C. Maxwell in 1865), *radiant energy* is made up of *particle-like* components known as *quanta.*

To prove his point, Planck calculated a value for the energy of each specific frequency (f), and he called it the *Planck constant,* denoted as "h" (Planck formula: E=hf). The energy of any photon can be calculated when the frequency of that photon is known.

Planck's constant is: h=6.63×10^{-34} Js.

In 1911, *Ernest Rutherford*, the New Zealand-born and English physicist, bombarded a piece of *gold foil* with a beam of positively charged *alpha particles* (alpha particles were discovered by Rutherford himself in 1899, and later, in 1907, they were identified as helium nucleus) and witnessed that some of the particles backscattered like they'd hit a wall (since alpha particles are positively charged, and the nucleus is also positively charged). He called this positively charged wall the *nucleus.* Later, he discovered that there are positively charged particles in the nucleus, and he called them protons. He also predicted neutrons, but he failed to find them. Rutherford is considered to be the *father of nuclear physics* for his many discoveries, such as the radioactive half-life (T1/2), the element radon, alpha particles, and protons. In the periodic table of elements, the element *rutherfordium-104* is named in his honor.

In 1913, one of Rutherford's students named *Niels Bohr*, a Danish physicist, proposed that electrons are not haphazardly floating in the atom as J.J. Thomson had suggested in 1897, but electrons are arranged in concentric circular orbits around the nucleus. His model is called the *planetary model*, like our solar system pattern.

The *Bohr model* can be summarized by four principles:

1. Electrons occupy only certain orbits that are stable.
2. Each orbit has its own energy.
3. Electrons can jump from one orbit to another. The one that goes to a higher orbit must *absorb* energy, and the one that falls to a lower orbit must *emit* energy (X-ray).
4. Emitted energy can be calculated, and it is called the *characteristic X-ray*.

In 1926, *Erwin Schrödinger,* an Austrian physicist, took Bohr's atom model one step further and mathematically showed that there are no exact paths of electrons orbiting the nucleus, but the nucleus is surrounded by an *electron cloud*.

Furthermore, Schrödinger and *Werner Heisenberg* both collaborated and mathematically discovered that there are no definite points within the atoms where the electrons can be found; however, the electrons can be located where the electron cloud is thickest.

Schrödinger's model also introduced the concept of *sub-energy levels* within the atom. His atomic model is known as the *quantum mechanical model*.

Until 1932, the atom was believed to be composed of a positively charged nucleus surrounded by negatively charged electron clouds. However, *James Chadwick,* also one of the students of Rutherford, bombarded *beryllium* (Be-4) atoms with *alpha particles* (α) and observed an unknown particle with a mass almost near the mass of a proton (p^+) and an electric charge of *neutral*. Later, this particle became known as a *neutron*. Neutrons were predicted by Ernest Rutherford around 1911, but he'd failed to discover them.

So, by 1932, all three constituents of an atom (proton, neutron, and electron) were discovered.

In the early 1900s, Max Planck's *quantum theory* revolutionized our understanding of *atomic* and *subatomic* processes, describing the *microworld,* and Albert Einstein's general theory of relativity (1915) revolutionized our understanding of space-time and gravity, describing the *macroworld*. Together, they constituted the fundamental theories of twentieth-century physics.

Later, other scientists such as Louis de Broglie, Paul Dirac, Enrico Fermi, Wolfgang Pauli, Satyendra Nath Bose, and Murray Gell-Mann advanced Max Planck's theory, leading to the era of *the Standard Model of elementary particles.*

Then physicists started to predict the existence of the *elementary particles,* such as *gluons*, *quarks*, and all the way to the discovery of *Higgs bosons*, which is one of the last and the most important discoveries of the elementary particles. The Higgs boson was predicted by Peter Higgs and George Zweig in 1964 and it was discovered forty-eight years later in 2012 by scientists working at the LHC (Large Hadron Collider) under the management of CERN (the European Organization for Nuclear Research).

CHAPTER SIX

"A rock pile ceases to be a rock pile
the moment a single man contemplates it,
bearing within him the image of a cathedral."
—Antoine de Saint-Exupéry

THE STANDARD MODEL OF ELEMENTARY PARTICLES (SM)

The *Standard Model* of elementary particles is the product of the relentless work of hundreds of determined scientists—those avant-garde thinkers, with their mathematical genius, God-given imaginations, and bold, gutsy personalities that allow them to dare to put their lifetime reputations on the line, in front of hundreds of unforgiving critical thinkers of the world, claiming guidance and complementary steps to the unfinished mathematical puzzles of our physical world.

They were the Columbuses of the uncharted world of quantum mechanics.

They were the *theoretical physicists* who passed the baton to the second relay runners, the *experimental physicists*—

the scientists who are known for their hard work, as they eat, breathe, and sleep to prove the righteousness of the mathematical world.

At last, in 2012, the final piece of the puzzle to the Standard Model of elementary particles was discovered. They called it the *Higgs boson* since it was predicted by Peter Higgs and some other scientists in 1964.

For now, the Standard Model of elementary particles is the bible of the world of quantum physics, although it is not a 100% solution to all the questions some scientists may have.

There are still many questions to be answered.

What is *dark matter*? Is this the matter holding the whole visible universe together, preventing it from falling apart? And if this is the case, how does it interact with normal matter?

Or what is *dark energy*? Is it the force causing the *expansion* of the universe to be faster and faster? Or what about the *graviton*? Is it the force responsible for mediating the force of gravity? Or why is there so much more *matter* in the universe as opposed to *antimatter*? There are many more unanswered questions that future scientists must wrestle with.

For the time being, the Standard Model is what we have, and we must cherish it and try to perfect it the best we can.

The Standard Model is so complicated, yet so simple! Let's dissect it!

WHAT IS AN ELEMENTARY PARTICLE?

The terms *elementary particles* and *fundamental points*

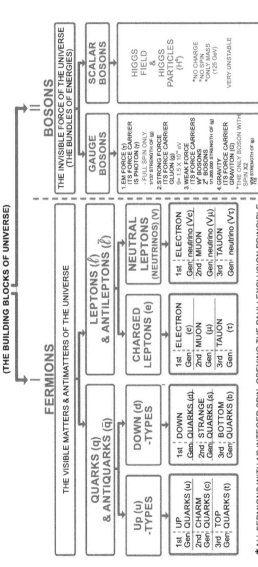

Fig. 13

mean that there are no known *components* or *substructures* for them. This was what, 2,470 years ago, in 450 B.C., Democritus, the Greek philosopher, said: "If you keep cutting an apple, there will be a point that the apple will be uncuttable," and he called that piece the "atom." However, our nuclear physicists, the Nobel Prize Laureates, should give him credit for what he said . . . and at the time he said it. Our scientists should turn a blind eye and say that he meant the elementary particles, and not the atom itself. I wish he was around today to see what he had started then, and where we are today! Oh boy!

Democritus was another "Bugsy" Siegel. Bugsy Siegel was the man who had a grandiose idea to build the first hotel-casino in Las Vegas, in the middle of nowhere. He built it and named it "The Flamingo." However, it unfortunately did not work out as he planned it, and subsequently, he lost his life because of it. Now we see how far his outrageous ambition has gone! Thank God for such people who see the world differently and push the envelope to make it possible for us to have the luxury and the comfort we enjoy today. The quotation by Antoine de Saint-Exupéry says it all: "A rock pile ceases to be a rock pile the moment a single man contemplates it, bearing within him the image of a cathedral."

Democritus and the other thinkers who followed in his footsteps were those who contemplated cathedrals, not rock piles.

According to the Standard Model, the visible universe is made of two substances: *matter* (fermions) and *energy* (bosons).

To get on with the task of learning about the Standard Model, it is a good idea that we first look at a bird's eye view of the table of the Standard Model to get a grasp of the whole subject that we are going to tackle. Then working on every detail will make more sense.

As we can see in Fig. 13, there are two main branches in the Standard Model.

I. *Fermions:* the matter, the body, or the atoms of the visible universe.

II. *Bosons:* the invisible bundles of energies that interact among the fermions, holding all the matter of the universe together.

FERMIONS

Fermions are the matter and antimatter particles of the visible universe. All fermions have a *half-integer spin* (spin is one of the three *intrinsic characteristics* of all elementary particles, the other two characteristics being the *charge* and the *mass*), and they all follow the *Pauli exclusion principle* (which will be explained later).

There are two types of fermions:
• Quarks and antiquarks
• Leptons and antileptons

THE FIRST BRANCH OF FERMIONS

Quarks (q) and Antiquarks (q⁻)

Quarks are one of the most important discoveries of the elementary particles, being the building blocks of all hadrons (including the most important ones, the protons and neutrons), held together with the power of the bosons.

Quarks were mathematically predicted by an American physicist, *Murray Gell-Mann,* in 1964, and were discovered in 1968 at Stanford Linear accelerator (SLAC).

There are two types of quarks:

A. The *up-type quarks (u):*

There are three weight classes of up-type quarks:

- up quarks (u)
- charm quarks (c)
- top quarks (t)

B. The *down-type quarks (d):*

There are three weight classes of down-type quarks:

- down quarks (d)
- strange quarks (s)
- bottom quarks (b).

There are also *three generations* of quarks, and in each generation, there are up-type and down-type quarks (Fig 13).

The *first generations* are the *up* quarks and *down* quarks, the *second generations* are the *charm* quarks and *strange* quarks, and the *third generations* are the *top* quarks and *bottom* quarks.

All three generations of quarks are practically the same; however, the lightest and most stable quarks are the first generation of quarks: the up quarks and down quarks. The second and third generations are heavier, and the heavier they get, the more *unstable* they become with a half-life of nanoseconds. They decay into some other lighter elementary particles.

Quarks have two types of charges:

First, there are the *fractional electrical charges.* All up-type quarks have *+2/3* charges, and all down-type quarks have *-1/3* electrical charges.

Second, the type of charges that quarks are assigned are called *color charges*; the arbitrary colors of *red, green,* and *blue* are given to each specific quark and antiquark. These are not real colors, but they are used for the sake of the conservation of the charges. The color charges are also assigned to gluons (one of the most important bosons).

Later, when we discuss *hadrons*, we will learn more about the color charges. The color charge system seems a bit odd and confusing. As *Richard Feynman*—one of the most flamboyant American theoretical physicists, Nobel laureate, and teacher—commented, saying: "It was an idiotic choice of words by those scientists."

The color charges of quarks and antiquarks help us to *match* the individual quarks together, and the fractional electrical charges of quarks and antiquarks help us to *calculate* their *electrical charges.* That is the way they came up with the (+1) charge of protons, the (s) charge for neutrons, and the (-1) electrical charge of electrons in an atom.

Each one of these quarks and antiquarks is denoted by one letter: quarks (q), antiquarks (q⁻), up quarks (u), charm quarks (c), top quarks (t), down quarks (d), strange quarks (s), and bottom quarks (b). All antiquarks are denoted by the same letter with a bar on top. All quarks and leptons are part of fermions, and all fermions have half-integer spins.

THE SECOND BRANCH OF FERMIONS

Leptons (ℓ) and Antileptons (ℓ̄) (Electron-like leptons)

This branch also is divided into two categories of:

A. *Charged leptons*

There are three weight classes of charged leptons:

- electrons (e-)
- muons (μ)
- Tau / tauons (τ)

B. *Neutral leptons or neutrinos (V)*

There are three weight classes of neutral leptons/ neutrinos:

- electron neutrinos (ve)
- muon neutrinos (vμ)
- tau / tauon neutrinos (vτ)

All leptons have their own electrical charges; therefore, they are not part of the color charge assignment. All charged leptons have (-1) electric charge, and neutral leptons / neutrinos (V) have zero charges (☉).

Electrons only combine with neutrinos with the help of the *weak interactions* (W and Z) to form electron neutrinos.

Like quarks, leptons come in three generations as well (Fig. 13).

The *first generations* are *electrons* (e-) and *electron neutrinos* (ve).

The *second generations* are *muons* (μ) and *muon neutrinos* (vμ).

And the *third generations* are *tau / tauons* (τ) and *tau / tauon neutrinos* (vτ).

Like different generations of quarks, different generations of leptons are also very similar to electrons, but the second and third generations of leptons (muons and tau / tauons) are much heavier than electrons (muons are 200 times the mass of an electron and tauons are 3,700 times the mass of an electron) and as a result, they are very unstable and decay in a few nanoseconds.

Unlike muons and tauons, electrons are one of the most stable fermions. However, the electron's mass is only 1/1836 the mass of a proton (protons and neutrons have approximately the same mass, while neutrons are a bit heavier). Like all elementary particles, electrons exhibit properties of both particles and waves.

NEUTRINOS

Neutral leptons—or neutrinos (V)—come in three generations of electron neutrinos (ve), muon neutrinos (vμ), and tau neutrinos (vτ).

Neutrinos are neutral elementary particles with a *mass close to zero* (1/1,000,000 mass of an electron's mass). Electrons, muons, and tauons can only combine with neutrinos, unlike the quarks that are able to combine with one another.

Neutrinos come from a few different sources:

1. Thermonuclear fusion in the *sun*.
2. Manmade sources such as *particle accelerators* and *nuclear reactors*.
3. The Earth's *natural decays of radioactive materials*.
4. High-speed protons of cosmic rays hitting the Earth's atmosphere, which creates *atmospheric neutrinos*.
5. The residual neutrinos from the Big bang, called *cosmic background neutrinos* (0.0004 eV).

Neutrinos are called the *ghostly particles* because they have zero charge, with almost zero mass, and travel at nearly the speed of light; therefore, they barely interact with matter. Trillions of them go through our bodies every second without being noticed.

Scientists are able to detect neutrinos that pass through our planet with a detector on the other side of the Earth.

BOSONS

Bosons are the second main branch of the Standard Model of elementary particles (SM).

The first group we discussed, the fermions, are the matter and antimatter of the visible universe, while the bosons are the *force carriers*, mediating the interactions among the fermions. Bosons hold the matter of the visible universe together. Unlike fermions that all have one half-integer spin and follow the Pauli exclusion principle, all bosons have *one full-integer spin* with the exceptions of the *graviton* having spin-2, and the *Higgs particles* having no spin. All bosons follow the *Bose-Einstein statistics*, meaning that bosons can *bunch up* in any *single state*. (These statistics and principles will be explained in more detail later.)

There are two types of bosons:
- Gauge bosons
- Scalar bosons

GAUGE BOSONS

For this type of boson, there are *four* fundamental *forces,* each with their own *force carriers:*

1. *EM force* and its force carrier, the *photon* (γ)

2. *Strong force* and its force carrier, the *gluon* (g)
3. *Weak force* and its force carriers, the *W*± bosons and the *Z*∅ bosons
4. *Gravity* and its force carrier, the *graviton (G)*

Before we start learning about bosons, I would like to present you with a very simple analogy in order to understand bosons easier.

As I understand, bosons are the forces *holding* the fermions, or the matter of the visible universe, together. There are four types of gauge bosons, each with specific holding power and different effective range for holding the elementary particles together. Without them, the universe would not exist as we see it today.

Let's imagine that we are going to build a house. *First,* we buy all kinds of materials we need, such as all types and sizes of lumber, tiles, doors, windows, etc. *Second,* we need to buy all different sizes of nails, screws, glues, and cement to hold all the lumber, doors, and windows together.

To start building the house, if we are putting together two pieces of 2x4 wood, we must use 16-gauge long nails, and for decorating the windows with delicate moldings, we need to use small ½-inch finish nails, otherwise it would not work. The small finish nails would not be able to hold the two pieces of 2x4 together, and the 16-gauge long nails would destroy our delicate moldings. As you see, different sizes, strengths, and lengths of materials are needed to build a house properly that will hold together.

To build hadrons from fermions, a specific boson is needed to hold a few specific elementary particles together.

One very important point is that bosons, unlike nails and screws, are invisible forces, and a good example of them is a refrigerator magnet holding a bunch of paperclips together with no visible attachments.

Bosons come in different *strengths* and *reaches*, just like different sizes of nails and screws. Therefore, each of the four invisible forces of bosons has a specific use in holding the different parts of an atom together. The specific use of each force will be discussed in detail.

There are two main aspects of each of these four bosons: *first*, how *strong* they are; and *second*, how *far* their effective range of grasp or hold is.

The Electromagnetic Force

Photon (y) is the force carrier of the EM force. The effective range of the EM force is infinite, and it is the second strongest boson in the universe. The magnitude or the holding power of the EM force is 1/137 of the strength of the *strong force*.

The Strong Force

Gluon (g) is the force carrier of the *strong force*, and it is the *strongest boson* in the universe. However, it has the shortest effective range of holding, which is only 0.8 fm (one femtometer is 10^{-15} meter). Gluons are responsible for holding different types of quarks together to form a hadron (such as, mesons, baryons, tetraquarks, etc.). They will be discussed fully in *hadron* formation (hadronization).

The Weak Force

The weak force is also effective in a short range of 1—3 fm. And its strength or holding power is 1/1,000,000 of the strength of the *strong force*. Unlike the other three fundamental forces (EM force, strong force, and the gravity), the weak force has two types of *force carriers*. The W$\pm$ bosons (with $\pm$ electrical charges) and z⊙ bosons (with neutral or zero charges).

The weak force, together with the *residual strong force,* are both responsible for holding the *protons* and *neutrons* together, to form the nucleus of an atom.

Weak force is also responsible for the processes during *radioactive decays* and in *nuclear fission.*

The Gravity

Gravity is the weakest of all four *gauge bosons*, with an infinite range. Its attraction power is thought to be $1/10^{30}$ of the *strong force*. In theories of quantum gravity, the graviton is a hypothetical force carrier of the gravity because it is so weak that it cannot be measured or quantized. If it exists, it is expected to be massless, because the force of the gravity is long range with a speed of light. Graviton is supposed to be a spin-2 gauge boson.

SCALAR BOSONS

In this category, there is only one *force,* and that is the *Higgs field,* and its quantum unit is *Higgs boson* (H). Out of the three *intrinsic properties* of every elementary particle, the Higgs bosons only possess the *mass property*. They lack *spin* and *electrical charges*. The Higgs boson's mass is about 125 GeV (10^9 eV). Because of the heavy weight, they are extremely unstable. The *life span* of a Higgs boson is about a six-trillionth of a second, or *as quick as a thought*. Therefore, they can travel only a few femtometers

① **Single Quark:** NOT FOUND IN NATURE.

② **Two Quarks combination / mesons:** ONE QUARK & ONE ANTIQUARK COMBINATION.

TYPES OF MESONS:

Ⓐ PIONS
(PI-mesons)
π^+ — $u\bar{d}$
π^- — $\bar{u}d$
π^0 — $u\bar{u}$ / $d\bar{d}$

Ⓑ KAONS
(K-mesons)
K^+ — $u\bar{s}$
K^0 — $d\bar{s}$ / $d\bar{s}$

Ⓒ ETA &
ETA prime
η — A COMBINATION of u, d, s QUARKS
η' — AND THEIR ANTIQUARKS

Ⓓ PHI-mesons
φ — $s\bar{s}$

③ **Triquarks/Baryons:**
(THREE QUARKS COMBINATION)
THE BARYON GENESIS

PROTONS "p" — udu
NEUTRONS "n" — dud

* A COMBINATION OF FOUR QUARKS AND MORE ARE CALLED "EXOTIC HADRON"

④ **Tetra-Quarks:** THIS COMBINATION HAS BEEN PREDICTED BUT NOT FOUND YET. IT IS A COMBINATION OF TWO QUARKS & TWO ANTIQUARKS.

⑤ **Penta-Quarks:** PREDICTED AND DISCOVERED. MADE OF FOUR QUARKS & ONE ANTIQUARK. THERE ARE TWO TYPES OF FORMATIONS:
Ⓐ MESON & BARYON MODEL (q-q̄ & q(red), q(green), q(blue));
Ⓑ FIVE QUARKS (BAG): q, q̄, q(red), q(green), q(blue) ALL MIXED WITH NO PATTERN.

⑥ **Hexa-Quarks:** PREDICTED BUT NOT FOUND YET.
(Six-Quarks) SIX QUARKS/SIX ANTIQUARKS COMBINATION OF ANY FLAVOR. IN THE SHAPE OF THE STAR OF DAVID.

HADRONS

HADRONIZATION: IT IS A PROCESS BY WHICH HADRONS ARE FORMED OUT OF QUARKS AND GLUONS. IT IS A BONDING SYSTEM.
HADRONS ARE COMPOSITE PARTICLES.
99% OF THE HADRON'S MASS IS CONTRIBUTED FROM THE GLUONS AND ONLY 1% OF IT IS FROM QUARKS.

Fig. 14

(fm) before they decay. *Higgs field* was theorized by two physicists named *Peter Higgs* and *Brout Englert* in 1964. And finally, forty-eight years later, on July 4, 2012, at LHC facilities under the supervision of CERN (The European Organization for Nuclear Research), the Higgs bosons were discovered. Higgs field was the answer to the question: where do elementary particles get their mass from?

Every single particle traveling with a speed under the speed of light gets its mass by disturbing or interacting with Higgs field. However, massless particles such as photons (y) and gravitons (G), which travel with the speed of light, don't disturb the Higgs field, and they don't gain any mass either.

My analogy of Higgs field is as follows: if your body is immersed in a pool of water and you are trying to walk or run underwater, you don't feel like a 170-pound person anymore, but much heavier, because your body is disturbing the water molecules, and water molecules are resisting to move around you.

Now we understand all the elementary particles and their intrinsic characteristics (mass, charge, and spin). However, these elementary particles cannot form an atom unless they go through a process called *hadronization*.

HADRONIZATION

Hadronization is a bonding system that creates *hadrons* out of quarks held together by bosons. Quarks cannot exist as an isolated elementary particle in nature due to their color confinement.

After the Big bang, when the quark-gluon plasma

cooled down, the processes of *hadronization* took place and hadrons were formed.

Hadrons are not part of elementary particles, but they are called the *composite particles,* such as composite bosons or composite fermions depending on their baryon numbers.

DIFFERENT TYPES OF HADRONS

Different types of hadrons are solely divided based on the number of quarks held together by gluons. That is, two quark combinations will form *mesons,* three quark combinations will form *baryons,* four quark combinations will form *tetraquarks,* and so forth (Fig. 14).

Gluons make up 99% of the mass of any hadron, and only 1% of the mass of the hadron is contributed by the *quarks*.

There are nearly 200 known hadrons in nature, but here, we will discuss only six of them.

1. *Single quarks:* Single quarks are not found in nature—they decay immediately.

2. *Two-quarks* combinations are called *mesons:* Mesons are composed of *two quarks of any kind* as long as one of them is a quark and the second one is an antiquark (i.e. u d⁻, u⁻ d, or s u⁻, s⁻ u, etc.).

There are at least *twenty types* of mesons that have been found, however, there are a few of them that are frequently mentioned. Here are four of them:

Pions / Pimeson / Yukawa potential (π):

Pions are not produced by *radioactive decays*; they are produced by *cosmic rays* or on Earth *in high-energy accelerators*. Pions have *zero spin* (integer spin) and are denoted by (π). There are three types of pions: π+ positive charge, π- negative charge, and π⚬ neutral charge.

Positive pions (π⁺) are composed of one up quark and one anti-down quark (u d⁻). *Negative pions* (π⁻) are composed of one anti-up quark and one down quark (u⁻ d). *Neutral pions* (π̊) is either composed of one up quark and one anti-up quark (u u⁻), or one down quark and one anti-down quark (d d⁻).

Charged pions and neutral pions decay differently. Charged pions decay into positive or negative muons and muon neutrinos. And neutral pions decay into gamma rays.

kaons / k-mesons (k):

With *zero spin,* there are also three charges of kaons: *positive kaons* (k⁺), made up of one up and one anti-strange (u s⁻); *negative kaons* (k̄), made up of one strange and one anti-up (s u⁻); and *neutral kaons* (k̊), made up of either down and anti-strange (d s⁻), or strange and anti-down quark (s d⁻). Positive and negative kaons decay into positive and negative muons and muon neutrinos. The neutral kaons will decay into gamma radiation.

Eta and Eta prime:

Denoted by (η and ή), both mesons are made of a mixture of up, down, and strange quarks and their antiquarks. Both have zero charges and decay into all charges of pions.

Phi mesons:

(Φ) is formed from a strange quark and a strange antiquark (S S⁻).

3. *Three-quarks* combinations (tri-quarks), which are called the *Baryons*. Two of the most important subatomic particles in any atom are protons and neutrons, and both are baryons. Protons (p) are composed of two *up quarks* and one *down quark* (uud). And neutrons (n) are composed of two down quarks and one up quark (ddu).

All quarks in the compositions of any types of hadrons are held together by gluons; gluons contribute 99% of the mass of the baryons, and quarks the remaining 1%. Baryons make most of the mass of the matter in the visible universe. Baryons, being in the fermion group (composite fermions), possess half-integer spin, and during their formation, they must follow the Pauli exclusion principle and the *neutral rules* of the *color charges*.

4. *Four-quarks* combinations (tetraquarks): A combination of *two quarks* and *two antiquarks* (qq and q⁻q⁻). They have been theorized but not discovered yet.

5. *Five-quarks* combination (pentaquarks): Different strands of pentaquarks have been

discovered in different labs: in Japan's LEPS facilities in 2003, and in CERN's LHCb in 2015 and 2019.

Pentaquarks are made of four quarks and one antiquark (qq^-, qqq). There are two models of quark combinations in pentaquarks.

First, the *meson-baryon model*, which there are two separate forms of, three quarks *baryon* (q red, q green, q blue), next to a two quarks *meson* (quark-antiquark). The *color confinement* for the three quarks are the three colors of red, green, and blue, to make the neutral color of white, and for the meson color combination, either of three colors with its own anti-color will make a neutral color (i.e. blue/anti-blue, red/anti-red, or green/anti-green).

Second, the five-quarks or the "bag" model. In this model, the number and types of the quarks are the same as the latter, but in this model, all quarks are glued together, yet considering the three- and two-quarks' combinations, and the color confinements.

6.   *Six-quarks* combination (hexaquarks): These have been predicted but not found yet. It is a combination of either six quarks or six antiquarks. It is in the shape of the Star of David, with two separate three-quarks baryons, one baryon on top of the other.

EXOTIC HADRONS

Hadrons with four or more quarks are called *exotic hadrons* and in their formations, their fractional *electrical charges* and *color charges* (red, green, and blue) must be fully considered.

The three colors of red, green, and blue in their hadron combinations must result in a *neutral white light*, and the two colors of the red and anti-red or green and anti-green or blue and anti-blue of mesons combinations (quark and antiquark) will also be a neutral white color, resulting in a proper color confinement.

Exotic hadrons, such as tetraquarks, pentaquarks, and hexaquarks, are classified as *baryons* or *mesons* depending on their baryon numbers.

To calculate the *baryon numbers* or the *lepton numbers*, we need to know the values of each *quark's* electric charge and each *lepton's* electric charge!

All *up-type quarks* have +2/3 electric charges.

All *down-type quarks* have -1/3 electric charges.

All *antiquarks* have the opposite charges of their quarks (-2/3 and +1/3).

All charged leptons have -1 electric charges.

All neutral leptons / *neutrinos* have zero electric charges.

All antileptons have the opposite charges of their leptons (+1), while they all have identical mass of their own leptons, such as electrons (-1), and positrons (+1).

Knowing these values, we can calculate the baryon numbers and the lepton numbers, which will determine if the *hadron* is a *baryon* or a *meson*.

Baryon numbers of baryons, mesons, and antibaryons

are as follows:

Baryon numbers of baryons=+1

Baryon numbers of mesons=0

Baryon numbers of antibaryons=-1

All antiparticles have the opposite charges of their same particles.

Note: the author believes that the name *baryon number* is a bit confusing and instead of calling them the baryon number, it should be called the *hadron numbers*. It will include all the hadrons (i.e. the hadron number of *baryons*, and the hadron number of the *mesons,* and the hadron number of *antibaryons*). However, this is only a suggestion, and I am sure the scientists who participated in naming this *value* have had their own reasons.

The formula for calculating the baryon number is:

$B=1/3(nq-nq^-)$

Where B is the baryon number, nq is the number of quarks, and the nq^- is the number of antiquarks.

And to calculate the lepton numbers:

$L=nl-nl^-$

Where L is the lepton number, nl is the number of leptons and nl^- is the number of antileptons.

CHAPTER SEVEN

"No reasonable definition of reality could be expected to
permit [Quantum Mechanics]."
—Albert Einstein

ELECTRON CONFIGURATION

Now that we understand *hadronization* and the formation
of the *baryons* (protons and neutrons)—the two constituents
of the atom's nucleus—the only missing component for the
formation of a complete atom is the *electron*.

Any atom in any element, regardless of being stable or
radioactive, *must* have the same number of protons and
electrons.

Now here comes the big warning! In an atom's world,
you don't want to be an electron!

Because, for an electron to find its way into an *orbital*
around a nucleus, there is a very sophisticated system
of *energy matching* that dictates and ushers each single
electron into a specific *orbital* within the *electron cloud*.

Each electron must and will go through a series of four

quantum numbers, and three *rules and principals* in order to be accepted and placed into the crazy roller-coaster world of an *electron cloud.*

I have a simple analogy that might help us understand what an electron must go through:

It is a Friday night. You are all alone, so you get dressed up and go out to see a play. You drive up to a theater. The valet takes your car away. You walk to the window and ask the person in charge for a ticket to see the show.

The first thing she is going to ask you is which section of the theater you would like to be seated, which is a very polite way of asking how much money you are willing to spend.

You, according to your *budget*, will buy a ticket.

Ticket in your hand, you approach the *first* person to let you inside the theater. Then, the *second* usher will look at your ticket and send you on your way to the balcony or any other section of the theater that your ticket indicates. The *third* usher will shine a flashlight onto your ticket and tell you that your seat is way in the front row of the balcony. You walk to the front of the balcony, where the *fourth* usher reads your seat number and shows you the seat that your ticket indicates. This time, he will follow you to make sure that you are seated next to a member of the opposite sex. No two women or two men can be seated next to each other, and that is the theater's policy!

Now, applying the roles of those *four ushers* in the theater to the four *quantum numbers* of the electron configuration, it will help us to understand how each *quantum number* ushers each single electron into the right

ELECTRON CONFIGURATION
(QUANTUM STATE OF ELECTRONS)

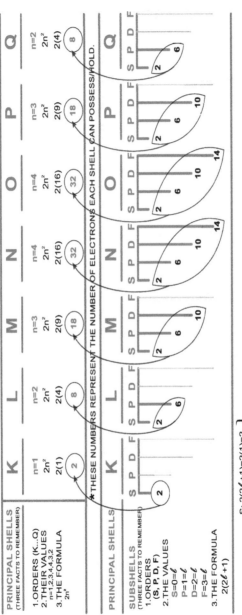

PRINCIPAL SHELLS (THREE FACTS TO REMEMBER)	K	L	M	N	O	P	Q
1. ORDERS (K...Q)							
2. THEIR VALUES $n=1,2,3,4,4,3,2$	$n=1$	$n=2$	$n=3$	$n=4$	$n=4$	$n=3$	$n=2$
	$2n^2$	$2n^2$	$2n^2$	$2n^2$	$2n^2$	$2n^2$	$2n^2$
3. THE FORMULA $2n^2$	$2(1)$	$2(4)$	$2(9)$	$2(16)$	$2(16)$	$2(9)$	$2(4)$
	2	8	18	32	32	18	8

*THESE NUMBERS REPRESENT THE NUMBER OF ELECTRONS EACH SHELL CAN POSSESS/HOLD.

PRINCIPAL SHELLS	K	L	M	N	O	P	Q
SUBSHELLS (THREE FACTS TO REMEMBER)	S P D F	S P D F	S P D F	S P D F	S P D F	S P D F	S P D F
1. ORDERS (S, P, D, F)	2	2 6	2 6 10	2 6 10 14	2 6 10 14	2 6 10	2 6
2. THE VALUES							

2. THE VALUES
$S=0=\ell$
$P=1=\ell$
$D=2=\ell$
$F=3=\ell$

3. THE FORMULA
$2(2\ell+1)$

S: $2(2\ell+1)=2(1)=2$
P: $2(2\ell+1)=2(3)=6$
D: $2(2\ell+1)=2(5)=10$
F: $2(2\ell+1)=2(7)=14$

SINCE EACH ORBITAL CONSISTS OF TWO ELECTRONS WITH OPPOSITE SPINS, THEN: S, SUBSHELL HAS ONE ORBITAL
P, SUBSHELL HAS THREE ORBITALS
D, SUBSHELL HAS FIVE ORBITALS
F, SUBSHELL HAS SEVEN ORBITALS

Fig. 15

orbital within the *electron cloud.* This system of control and electron placement within any atom is called *the electron configuration* (Fig. 15).

THE FOUR QUANTUM NUMBERS

The electron configuration consists of *four quantum numbers* (like the four ushers in the theater).

1. The Principal quantum number (n).
2. The Azimuthal quantum number (ℓ). It is also called the *orbital angular momentum* quantum number.
3. The Magnetic quantum number (ml).
4. The Spin quantum number (ms).

When the electrons are in their right orbitals, then the *quantum state* of the atom is completed. Furthermore, there are *three principles / rules* that complement the guidance of those four quantum numbers.

The three additional principles/rules are:

- *The Pauli exclusion principle*
- *The Hund's rule*
- *The Aufbau / Wiswesser rule*

At this point, it would be a good idea to look at Fig. 15. It will give you a mental picture of the whole processes of electron configuration, making it easier to follow each step involved. You can learn 80% of the electron configuration by only studying the Fig.15.

The good news is, if you understand the calculations of the first two *quantum numbers*, the rest is easy.

THE PRINCIPAL QUANTUM NUMBER (n)

The first quantum number explains the seven *principal shells* of an atom.

In order for us to understand how many electrons each principle shell can hold, we must go through three steps. *First*, the *order* of principal shells. *Second*, the *value* of each principal shell. *Third*, the *formula* being used to calculate the number of electrons for each principal shell.

First step is the *order* of these principal shells that are denoted as: K, L, M, N, O, P, Q.

Second step is the *values* (denoted as "n") of these principal shells, which is different for each principal shell.

The value of K-shell, n=**1**; for L-shell, n=**2**; for M-shell, n=**3**; for N-shell, n=**4**; for O-shell, n=**4**; for P-shell, n=**3**; for Q-shell, n=**2**.

The *third* step that we must memorize is the *formula* $(2n^2)$.

Now, by knowing the *order*, *value* and the *formula*, we can calculate the number of electrons that each principal shell can *hold*.

Let's do some examples:

How many electrons can the M-shell hold?

The order is M, the value is n=3, and the formula is $(2n^2)$.

2(9)=18. So, the principal M-shell can hold only 18 electrons and no more.

Now let's solve for principal O-shell.

The order is O, the value is, n=4, the formula is $(2n^2)$. Then, 2(16)=32.

If we solve the problem for all principal shells, we will

come up with these numbers.

K-Shell can house 2 electrons.

L-Shell can house 8 electrons.

M-Shell can house 18 electrons.

N-Shell can house 32 electrons.

O-Shell can house 32 electrons.

P-Shell can house 18 electrons.

Q-Shell can house 8 electrons.

Therefore, the heaviest element in the periodic table of elements, with all its principal shells completely full, can hold ONLY up to 118 electrons.

Remember! In every atom / element, the number of electrons and protons are ALWAYS even.

THE AZIMUTHAL QUANTUM NUMBER (ℓ)

(The orbital angular momentum)

This quantum number represents the *subshells* (S, P, D, F) and their *orbitals* (each orbital can only hold two electrons with opposite spins) within each subshell. As shown in Fig. 15, each principal shell consists of four subshells.

To solve for this *quantum number*, we must follow the same three steps as we did for the principal shells: the *order*, the *value*, and the *formula*.

The *order* for azimuthal quantum number is: S, P, D, F.

The *values* (ℓ) for each subshell are: S=0, P=1, D=2, F=3.

The third piece of information that helps us to find out the number of electrons within each subshell is the *formula*: $2(2\ell+1)$.

Azimuthal quantum number represents the atom's

subshells and each subshell's number of *orbitals.*

Just like the principal shells that can hold certain numbers of electrons, the total number of electrons that one subshell or two subshells or three subshells or all four subshells can hold, cannot exceed the number of electrons that their own principal shell can hold (Fig. 15).

According to the periodic table of elements, the element hydrogen has only one electron, and the last element in the table of elements, *oganesson*, can hold 118 electrons.

Earlier, we did solve for the maximum number of electrons that each of the seven principal shells can hold.

Now the Azimuthal quantum number will show us how many electrons, or *pairs of electrons* (orbitals), each subshell can hold.

To start calculations of the Azimuthal quantum number:

First, we must consider the *order* of these subshells: S, P, D, F.

Second, the *value* of each subshell: S=0, P=1, D=2, F=3

Third, the *formula*: $2(2\ell+1)$ will tell us how many electrons each subshell can fill. And subsequently, by dividing the number of electrons by two in each subshell, we can tell how many *orbitals* there are in that subshell.

In this formula, "ℓ" is equal to the value of each subshell.

Now we can solve for each subshell.

To find out how many electrons each subshell can hold, first, it is important to know what *principal shell* the *subshell* is under or belongs to, and second, how many electrons that principal shell is allowed to house or hold.

When the number of electrons in subshells equals the number of the electrons in the principal shell, the calculation is over and we will proceed to the next *principal shell* and its subshells.

Let's solve for: "S" subshell in "K" principal shell.

Since this portion of the electron configuration is kind of confusing, it will help if we refer to Fig. 15 more frequently.

First, we must keep in mind that K-shell can only hold *two* electrons.

Solving for "S" subshell under the K-principal shell,

$\ell=0$, and the formula is: $2(2\ell+1)$, then,

$2(0+1)$ the number of electrons for "S" subshell in the K-principal shell or any other principal shell will be equal to 2.

However, since the maximum number of electrons in principal K-shell is 2, then we are done with all subshells within the principal K-shell, and we must move on to next principal shell.

Next problem. Solving for the number of electrons in subshell D in principal shell M. We must consider that principal shell M can only hold 18 electrons, and we must stop calculations when the number of electrons in subshells within the principal shell reaches 18.

The *value* of subshell D is equal to two ($\ell=2$). And applying it to the formula, $2(2\ell+1)$, the result will be:

$2(4+1)=10$.

Now, let's pause for a minute and account for all the electrons that have been filling all the subshells until this point (Fig. 15).

Since we know that the capacity of the M-principal shell is only 18 electrons, then no more electrons will be permitted in the principal shell M, because the subshell

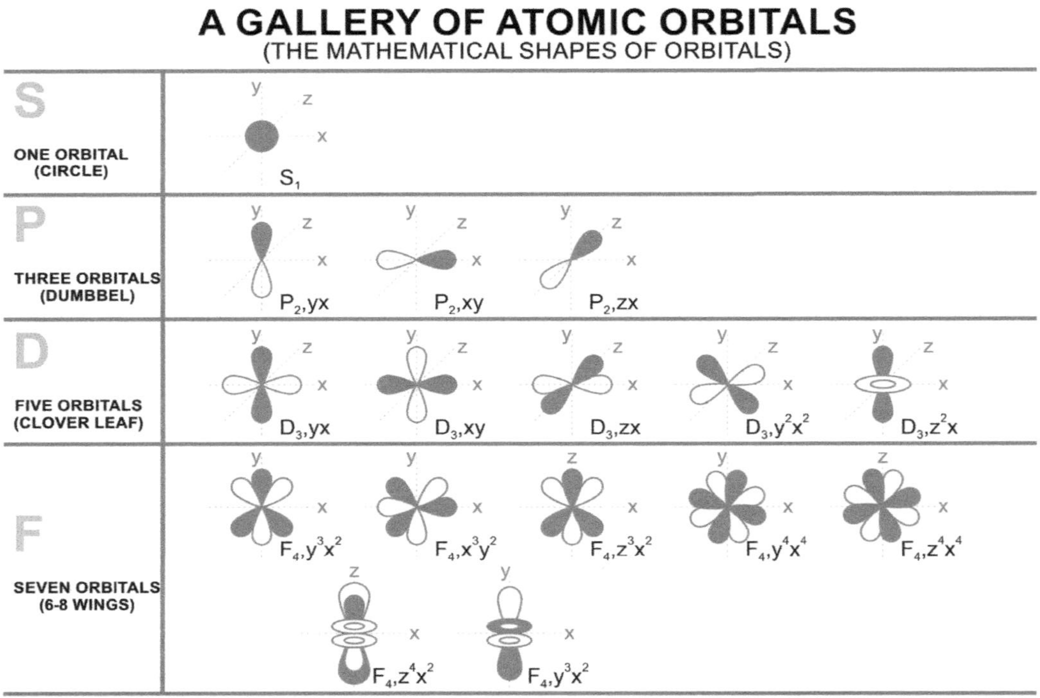

A GALLERY OF ATOMIC ORBITALS
(THE MATHEMATICAL SHAPES OF ORBITALS)

S — ONE ORBITAL (CIRCLE)

S_1

P — THREE ORBITALS (DUMBBEL)

P_2, yx P_2, xy P_2, zx

D — FIVE ORBITALS (CLOVER LEAF)

D_3, yx D_3, xy D_3, zx D_3, y^2x^2 D_3, z^2x

F — SEVEN ORBITALS (6-8 WINGS)

F_4, y^3x^2 F_4, x^3y^2 F_4, z^3x^2 F_4, y^4x^4 F_4, z^4x

F_4, z^4z^2 F_4, y^3x^2

Fig. 16

S is filled with 2 electrons, the subshell P is filled with 6 electrons, and the subshell D is filled with 10 electrons, which all bring it up to 18 electrons; that is all the electrons the principal shell M can hold. And that would be the end of the calculations for the principal M-shell.

$(2+6=10=18)$.

The calculations for the rest of the subshells in all principal shells reveal each subshell can hold a certain number of electrons.

Since, each *orbital* is filled with *two* electrons with opposite spins, dividing the number of electrons in each *subshell* will give us the number of *orbitals* in each *subshell*.

Subshell S can only hold 2 electrons (one orbital), regardless of being in any principal shell.

Subshell P can only hold 6 electrons (three orbitals) regardless of being in any principal shell.

Subshell D can only hold 10 electrons (five orbitals) regardless of being in any principal shell.

Subshell F can only hold 14 electrons (seven orbitals) regardless of being in any principal shell.

These capacities coordinate with the number of electrons each *principal shell* can handle.

NOW, WHAT IS AN ORBITAL?

An orbital is a two-electron unit with opposite spins, within each subshells of S, P, D, or F. Electrons are the only constituents of the orbitals. The history of electrons started with *J.J. Thomson,* the English scientist, in 1897. He discovered *electrons*, and he believed they were floating in a positively charged, jellylike sphere, which he called *the plum pudding model.*

Sixteen years later, in 1913, *Niels Bohr,* a Danish physicist, suggested that electrons were orbiting the nucleus with specific energy for each orbit, like planets going around the sun, and he called it *the planetary model.*

Finally, in 1927, *Erwin Schrödinger,* the Austrian-Irish physicist, took Bohr's atom model one step further and mathematically proved that there are no perfect orbits around the nucleus, but the nucleus is surrounded by a cloud of electrons, which he aptly called *electron clouds.*

Later, Schrödinger, with the help of a German theoretical physicist named *Werner Heisenberg,* mathematically determined that the regions in which the electrons would be most likely to be found are where the electron cloud is densest.

Remember that in our theater analogy, the theater policy for the sitting arrangement was one man and one woman, not two men or two women sitting next to each other.

In quantum physics, the Pauli exclusion principle dictates the arrangement of two electrons in each orbital.

The Pauli exclusion principle states that no two *fermions* (electrons, leptons, and quarks), with the same *one-half spin,* can occupy the same orbital. Thus, each orbital is made of two electrons each with opposite *spins:* spin up, +1/2 spin and spin down, -1/2 spin.

THE MAGNETIC QUANTUM NUMBER (ml)

Magnetic quantum number distinguishes the availability of the *orbitals within the subshells* and distributes the electrons into each orbital.

Magnetic quantum number functions under the three principles: Pauli exclusion principle, Hund's rule, and Aufbau rule (they will be discussed in detail).

THE SPIN QUANTUM NUMBER (ms)

Spin quantum number describes the *energy*, *shape,* and *orientation* of the orbitals.

The shapes of the orbitals are defined mathematically, and it is kind of hard to imagine how the orbitals are piled up one on top of the other in an electron cloud.

If the movement of electrons within the electron cloud is not crazy enough, some electrons move in and out of their orbitals by emitting or absorbing energy. The orbital shapes are shown in "A Gallery of Atomic Orbitals" (Fig. 16).

The best way I can describe an *orbital gallery* is by imagining an onion with seven layers, with each layer possessing four thinner sub-layers (subshell). Now imagine that each *subshell* has its own *shape,* regardless of being in any principal shell. Subshell S is shaped *circular*, subshell P is shaped like a *dumbbell*, subshell D is *clover leaf* shape, and the subshell F is a *6- to 8-petal* flower shape.

All these shapes will be sitting one on top of the other like a *three-dimensional puzzle*.

The first principal K-shell, containing only one subshell S, is the shape of a *circle*.

The second principal L-shell would be in the shape of one circle and one dumbbell sitting one on top of the other. The third principal M-shell would be in the shape of one circle, one dumbbell, and one clover leaf sitting one on top of the other, and so forth (Fig. 16).

For a complete periodic table of elements, there would be 59 orbitals, in four shapes of circle, dumbbell, clover leaf, and 6- to 8-petal flower, sitting one on top of the other.

All these orbitals have their own energies and shapes, orbiting around the nucleus in a cloud-like fashion.

THE THREE PRINCIPLES / RULES

Now we are done with all four quantum numbers, and in order to complete the whole processes of the electron configuration, the *three* mandatory principles and rules must be followed by every single electron in order to end up in their right *quantum state*.

THE PAULI EXCLUSION PRINCIPLE

The principle of Pauli Exclusion prohibits two or more identical elementary particles of fermions (having half-integer spin) to occupy the same *quantum state*. In an orbital, the electrons must have opposite spins (-1/2 spin and +1/2 spin).

Also, how these electron distributions take place is supported and described by *Fermi-Dirac statistics*.

THE HUND'S RULE

In a subshell with multiple orbitals, electrons will occupy each orbital *singly* before occupying any of the orbitals *doubly* (i.e. it is known that subshell S has one orbital, subshell P has three orbitals, subshell D has five orbitals, and the subshell F has seven orbitals). To follow the Hund's rule in filling of each orbital, if the orbitals in subshell D are being filled, all five orbitals will get one electron in each orbital before getting their second electrons.

And the second electron occupying any orbital must follow *the Pauli exclusion principle*.

THE AUFBAU (BUILD-UP) PRINCIPLE

In the ground state of an atom, electrons fill all atomic orbitals of the *lowest available energy level* before occupying the orbitals of a *higher level*.

Note: *Wiswesser* table (Fig. 17) is a visual tool to help us follow the electrons filling the *orbitals* of every element within the periodic table of elements.

NOTE:

THIS TABLE HAS BEEN MODIFIED BY THE AUTHOR FOR EASIER MEMORIZATION; HOWEVER, THE ORDER OF FILLINGS OF THE ORBITALS FOLLOWS THE PRINCIPLES OF HUND & AUFBAU.

* THE 8TH PRINCIPAL SHELL OF "R" AND ITS SUBSHELL "S8" IS A PREDICTION OF TWO MORE ELEMENTS WHICH WILL BE ADDED TO THE PERIODIC TABLE OF ELEMENTS WITH **Z=119 & Z=120**

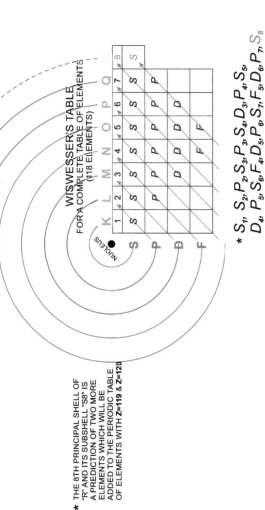

WISWESSER'S TABLE
FOR A COMPLETE TABLE OF ELEMENTS
(118 ELEMENTS)

$$* S_1, S_2, P_2, S_3, P_3, S_4, D_3, P_4, S_5,$$
$$D_4, P_5, S_6, F_4, D_5, P_6, S_7, F_5, D_6, P_7, S_8$$

Fig. 17

CHAPTER EIGHT

"If you don't do the best with what you have,
you could never have done better with what you could
have had!"
—Ernest Rutherford

STABLE ELEMENTS VS. RADIOACTIVE ELEMENTS

After the formation of the atoms, different *elements* are formed. Each *element* is a substance made of *the same type of atoms,* meaning atoms with the same atomic number (same number of protons in their nuclei). There are four forms of elements in nature: *solid, liquid, gas,* and *plasma.* Each one of these forms of matter / elements—beside the plasma state—can be in either a *stable* state (calm, quiet, content), or in an *unstable* or *radioactive* state.

I have a simple analogy for the stable atoms becoming unstable or radioactive and vice versa.

If you have a glass of water at room temperature, a couple of ice cubes, and a kettle of boiling water, the room temperature water is the *stable* form of water, and both the boiling water and the ice cubes are the *unstable* forms of water. In the case of the boiling water, the room temperature water has been given some extra heat that caused the water to boil. The boiling water must *give off* all the *heat* that was given to it in order to become a stable room temperature water again. And in the case of the ice cubes, some quantity of heat has been *taken away* from the stable room temperature water in order to make ice cubes. The ice cubes must gain the heat that has been taken away from them in order to become a stable room temperature water again.

This analogy shows if any kind of energy (in the case of water the *heat*) is given or taken away from the atoms of any element, it will become *unstable* or *radioactive*. For a radioactive element to go back to their stable state, they must lose or gain the same amount of energy that was given or taken away from them.

In the case of an atom, earlier, we learned that the only subatomic particles that can cause an atom to stay stable or become unstable (radioactive) are within the atom's nucleus. There are a few modes of decay that will change the number of nucleons within the nucleus, which will cause a stable atom to become radioactive. These processes can be done in cyclotrons, particle accelerators, and nuclear power plants, by *taking* or *adding* a proton or a neutron from their nucleus. These are the *manmade* radioactive materials.

However, not all radioactive elements on Earth are manmade. There are many radioactive elements that are

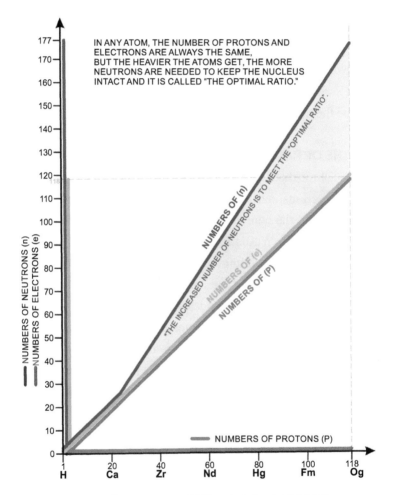

IN ANY ATOM, THE NUMBER OF PROTONS AND ELECTRONS ARE ALWAYS THE SAME, BUT THE HEAVIER THE ATOMS GET, THE MORE NEUTRONS ARE NEEDED TO KEEP THE NUCLEUS INTACT AND IT IS CALLED "THE OPTIMAL RATIO."

NUMBERS OF (n)

"THE INCREASED NUMBER OF NEUTRONS IS TO MEET THE "OPTIMAL RATIO".

NUMBERS OF (e)

NUMBERS OF (P)

NUMBERS OF PROTONS (P)

NUMBERS OF NEUTRONS (n)
NUMBERS OF ELECTRONS (e)

THE "OPTIMAL RATIOS" OF PROTONS AND NEUTRONS IN AN ATOM FOR THE ENTIRE TABLE OF ELEMENTS

Fig. 18

created by nature itself. We have about 3,800 radioactive isotopes on Earth. About 200 of them are being used for different purposes on a daily basis in hospitals and other industries. *Radioisotopes* of the same element are the elements with the same atomic number / proton number (in order to stay the same *element*) and the same number of electrons (in order to keep the same *chemical properties*). However, the only subatomic particle that changes would be the number of *neutrons*.

THE OPTIMAL RATIO OF PROTONS TO NEUTRONS

Throughout the periodic table of elements, the number of electrons and protons are the same for each element. However, the number of neutrons in heavier elements will increase. The reason for that is because, as the number of positively charged protons increases inside the nucleus, the repulsive force of *coulomb repulsion* will also increase, which can rip apart the nucleus. The only way to prevent that from happening is by creating a neutral buffer zone between the protons. And that is done by increasing the number of neutrally charged neutrons inside the nucleus.

Fig. 18 shows that there is an *optimal ratio* of the number of neutrons to protons in order to keep the heavier elements intact and in a stable state.

Now, the question is: are all radioisotopes from the same element the same? The answer is a big no! All different radioisotopes have different energies (eV) and different physical half-lives (Tp). However, they all have the same *chemical properties*, since the number of electrons in all of them are the same.

T1/2: SPONTANEOUS RANDOM DECAY

A Sample of:

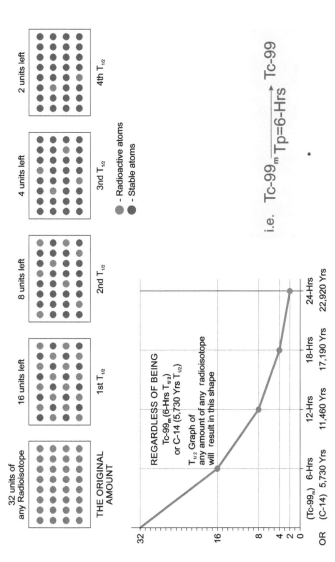

32 units of any Radioisotope

THE ORIGINAL AMOUNT

16 units left
1st T$_{1/2}$

8 units left
2nd T$_{1/2}$

4 units left
3nd T$_{1/2}$

2 units left
4th T$_{1/2}$

- Radioactive atoms
- Stable atoms

REGARDLESS OF BEING
Tc-99$_m$ (6-Hrs T$_{1/2}$)
or C-14 (5,730 Yrs T$_{1/2}$)

T$_{1/2}$ Graph of
any amount of any radioisotope
will result in this shape

(Tc-99$_m$)	6-Hrs	12-Hrs	18-Hrs	24-Hrs
OR (C-14)	5,730 Yrs	11,460 Yrs	17,190 Yrs	22,920 Yrs

i.e. Tc-99$_m$ $\xrightarrow{\text{Tp=6-Hrs}}$ Tc-99

*

Fig. 19

Earlier, we defined the electron volt (eV), and now we need to talk about the half-life (Tp / T1/2).

THE PHYSICAL HALF-LIFE (TP / T1/2)

The physical half-life is an interesting phenomenon that was discovered by *Ernest Rutherford,* a theoretical physicist, in 1907. *Physical half-life (Tp)* by definition is *the time that is required for an atomic nucleus of a radioactive sample to decay to half of its initial value.* The half-lives (Tp) of various radioisotopes are different and they range from one nanosecond (10^{-9}) up to billions of years (10^{+9}). It is a unique value that it is only affected by *time* and nothing else; even *temperature* and or *pressure* have no effect on it. Physical half-life is used in many industries, including anthropology to calculate the date of archaeological artifacts, and in hospital settings for the measurement of radiopharmaceuticals for the purposes of diagnostic and therapeutic procedures.

Physical half-life (Tp) is a *spontaneous* and *random* process in which the unstable radioactive atoms will transform into a stable atom. However, the atoms do not disappear and will remain in the sample as a stable atom (Fig. 19).

Fig. 19 shows the random decay of the atomic nuclei of a radioisotope sample and how exactly T1/2 works.

In a professional setting, to account for decay during each physical half-life (Tp), the professionals need a *reference chart* to assist them with the rate of decay for different radioisotopes. The reference chart is called the *isotope decay chart*. It shows the amount of radioactivity left after each Tp.

The isotope decay chart has three columns. The first column shows the *name* of all radioisotopes, the second column is the *elapsed time* for that radioisotope, and the third column is called the *decay factors,* which is a column of decimal point numbers (%) of the radioactivity left after each indicated time.

In an isotope decay chart, the percentages start at the top with the value of 0.9837, and at the bottom with the value of 0.0045. Of course, these percentages are different for different radioisotopes, and for that there are different windows showing the names of different isotopes.

Let's say in our lab, we measure the amount of Tc-99m in a dose calibrator, and it shows that we have 100 mci of Tc-99m at 8 a.m. We can calculate the amount of the activities left for any amount of time before or after our present time calculation.

Let's say we would like to know how much of our present activity would be left five hours after or five hours earlier.

Let's say that on the *isotope decay chart,* the *decay factor* for Tc-99m at five hours post calculation shows 0.5618 remaining.

If we multiply our *initial activity* (100 mci) by the decay factor of 0.5618, we will have 56.18 mci Tc-99m remaining five hours later. On the other hand, let's say we have 100 mci Tc-99m and we want to know how many millicuries we had five hours earlier. We will simply divide our 100 mci by the decay factor (0.5618), which it will give us 178.0 mci, the activity that we had five hours earlier that day.

DIFFERENT MODES OF RADIOACTIVE DECAYS

Now, let's see what kinds of elements we have here on Earth and how they transform from one element to the other through a process called *radioactive decay*.

On the periodic table of elements, from the hydrogen (H-1) up to the element *bismuth* (Bi-83), they all have their own naturally formed *stable* and *unstable* (radioactive) atoms, whereas from the element *polonium* (Po-84) up to the element uranium (U-92), they all exist only in natural *radioactive* forms.

Furthermore, from the element *uranium* (U-92) to the last element on the periodic table, which is the element *oganesson* (Og-118), they all are synthetic, manmade in our laboratories, cyclotrons, power plants, and nuclear reactors.

We have 118 elements in our periodic table of elements, and this number of elements are just the tip of an iceberg. There are another 3,800 different types of *radioisotopes* belonging to these 118 elements that are not shown in the table of elements. Each of these elements has somewhere between one to tens of radioisotopes of their own.

Decay processes are initiated from the element's *nucleus*, rather than any changes in their orbital electron(s).

There are many *modes* of radioactive decays that can change any *stable* atom into an *unstable* atom (radioisotope), and vice versa.

There are two modes of decay: I and II.

I. *The internal modes* of radioactive decays:

- Neutron decay

- Electron capture
- Internal conversion
- Alpha particles decay

II. *The external modes* of radioactive decays:

  A. Photo electric interactions

- Photoelectric effect
- Compton effect
- Inverse Compton effect

  B. Electron scattering

- The Bremsstrahlung radiation
- The Excited state
- The Elastic collision

THE INTERNAL MODES OF RADIOACTIVE DECAYS

There are four types of the internal modes of radioactive decays:

Neutron (n) decay (Negative beta decay)

In a *neutron-rich atomic nucleus*, a neutron will decay into a *proton*, an *electron (negative beta decay),* and an *antineutrino*.

When this happens, the proton will remain in the nucleus, transforming the nucleus into a new element with a higher atomic number (Z+1), while the electron (as a beta minus decay) and the antineutrino will leave the nucleus.

Beta particles travel around 1-4 meters with an energy of 0.511 MeV. The best shielding materials for beta particles are the low *atomic mass* materials such as water, wood, or plastic.

Beta emitter radioisotopes, such as I-131, are used in

nuclear medicine for noninvasive therapy of hyperthyroidism and thyroid cancer.

Electron capture (εc) (Positive beta decay or positron decay)

This type of decay takes place in *proton-rich atomic nuclei* when one of the atom's own orbiting electrons (usually from the K or L principal shells) is sucked inside the nucleus to be absorbed by one of the positively charged protons.

In this type of decay, a *proton* and an *electron* will combine, resulting in the formation of a *neutron* and an *electron neutrino (Ve)*.

However, a *positron decay* will take place if the energy difference between the parent atom and the daughter atom is bigger than 1.022 MeV.

If the difference is less than one mega electron volt, there would be no positron decay.

The positron (a positive beta decay), and the electron neutrino will be emitted from the nucleus while the loss of one proton in the nucleus will result in the conversion of an element to a new element (a daughter element) with an atomic number of one less proton (Z-1). But the atomic mass (P+N) will remain the same because the nucleus loses one proton while gaining one neutron.

Positron emitting radioisotopes, such as F-18, are used in Positron Emission Tomography (PET) scanning for production of the radiopharmaceutical F-18-FDG to be used for brain imaging in epilepsy and Alzheimer disease.

After the completion of the *electron capture*, the

processes of filling up of the *electron hole* by a higher orbital electron will take place, producing the emission of a *characteristic X-ray* and possibly emission of an *Auger electron.*

Furthermore, the *heavier* the element, the more episodes of *characteristic X-rays*, and the *lighter* the element, the more episodes of *Auger electrons* will be observed.

Internal Conversion (y)

It is a radioactive decay process, wherein an *excited nucleus*, by emission of a *photon* (γ), will cause the ejection of one of its *orbital electrons* out of the atom. The ejected electron is called the *conversion electron.* (If the ejection of the electron is caused by the atom's orbital X-ray, then it is called the *Auger electron.*) During *internal conversion*, the atomic number (Z) will not change, and no elemental *transmutation* will take place (no changing from one element to the other).

Alpha Particles (α) Decay

Alpha particles are a chunk of *subatomic particles* ejected out of a nucleus. Alpha particles consist of two protons and two neutrons. Since this combination resembles the nucleus of a helium atom (He), they are also known as the *helium atom's nucleus.*

Alpha particles are by far the most common form of *cluster decay,* which occurs in the heaviest nuclei, transforming an element by spontaneous fission-type processes. For example, U-238 decays into Th-234 (thorium-234) plus an alpha particle (atomic mass of 4: two protons and two

EXTERNAL MODES OF DECAY

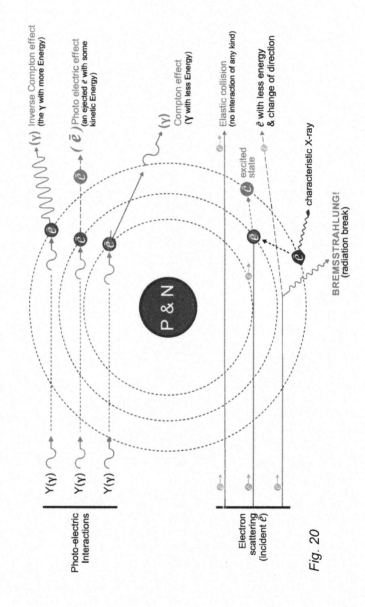

Fig. 20

neutrons).

The alpha particles' range of travel is only 4 centimeters or 1.5 inches, because they are *very heavy*, have relatively *low velocity*, and have an *electric charge of +2 e,* which causes them to lose energy fast and get stopped even by a piece of paper or human skin.

Furthermore, approximately 99% of helium produced on Earth is the result of the alpha particles decay of underground deposits of minerals containing uranium or thorium. The helium gas is brought to the surface as a byproduct of natural gas production.

When alpha particles are out of the nucleus, they start ionizing other atoms until they lose enough energy to slow down and capture two electrons to become a *helium atom.* Any element going through alpha particle decay will be transformed to an element with 2 protons less (Z-2).

An overview of all different internal modes of radioactive decay shows that all changes originate from the *nucleus* and the two particles of *alpha* and *beta*—as well as *gamma*—radiation are the balancing factors, contributing to stability or instability of an atom.

There are four conditions causing the emission of these balancing factors:

1. *The neutron decay* (n) within a nucleus will produce negative beta (β^-).

2. *The Electron capture* (εc) decay within a nucleus will produce positive beta or positron (β^+).

3. *The Internal conversion* decay within an excited nucleus will produce gamma radiation (γ).

4. *Alpha particles* decay, the heaviest nuclei going

through *cluster decay*, will eject a chunk of subatomic particles, such as alpha particles (α).

THE EXTERNAL MODES OF RADIOACTIVE DECAY

The incident elementary particle causing any atom going into a decaying mode could be either an incident *photon* or a *charged particle* such as an *electron.*

If the cause of decay is an *incident gamma ray* or *photon*, the phenomenon is called *photoelectric interactions.*

If an *incident electron* is the cause of decay, it is called *electron scattering.*

Each mode has its own types.

A. Photoelectric Interactions

If an *incident photon* interacts with an atom's electron cloud, one of these three decay modes will take place.
- Photoelectric effect
- Compton effect
- Inverse Compton effect

Photoelectric effect will take place when the energy of the incident *photon* is higher than the *binding energy* of the orbiting electron, and it will cause the ejection of the electron from the atom. The extra energy of the photon will be transferred to the ejected electron as *kinetic energy.* The ejected electron is called an *Auger electron.* And as a necessity of balancing the electrical charges at the orbital level, an upper-level electron will descend to the lower orbit to fill up the *electron hole* created by the ejected electron. Since the upper-level electrons possess more energy than the lower-level electrons, the extra energy will be released from the atom and it is called *characteristic X-ray.*

Compton effect: In this case, the photon's energy is less than the electron's binding energy, and when they collide, the photon will lose some of its energy to the electron and that will cause a change in the direction of the incident photon. As a result, at the point where the photon's direction of motion was changed, an X-ray or gamma radiation will be emitted. The photon will exit the atom with less energy.

Inverse Compton Effect: This phenomenon is the opposite of the *Compton effect*, where the *incident photon* will steal some of the energy from the atom's orbital electron and leave the atom with more energy.

B. Electron Scattering

Electron scattering takes place when an *incident electron* (charged particle) interacts with an atom's electron or positively charged nucleus. As a result, one of these three situations might take place:

- Bremsstrahlung
- Excited state
- Elastic collision

Bremsstrahlung: In this interaction, the incident electron will be pulled toward the *positively charged nucleus* of the atom. In this case, the nucleus is not powerful enough to *capture* the electron, and the electron will escape. However, the incident electron will lose some of its energies at the point where there was a pull by the nucleus, and that causes a change of its *direction of motion*. At the point of the change of direction, the lost energy will manifest itself as a *photon radiation,* which is called *Bremsstrahlung*.

The Bremsstrahlung radiation is important, as this is the method of generating X-ray in an X-ray tube or in linear accelerators.

Excited state: In this type of interaction, an incident electron and one of the atom's electrons will collide, wherein the *incident electron* will lose all its energy to the orbital electron, sending it to a higher shell / orbit, creating an *Excited state*.

And again, in this situation, the processes of filling the vacant space (*electron hole*) by another electron from a higher orbit will take place and the extra energy of the higher shell electron will be released as a characteristic X-ray from the atom.

Elastic collision: In this case, the incident electron will pass through the atom with no interaction with any orbital electrons. Therefore, no loss of energy, no change of direction, and no gamma radiation emission will take place (Fig. 20).

CHAPTER NINE

"Speak softly and carry a big stick, you will go far."
—West African proverb

FUSION AND FISSION

Fusion and Fission are certainly two of the most important atomic phenomena in the history of the universe.

NUCLEAR FUSION

Nuclear fusion was the beginning of the universe as we know it today. It is a process by which two hydrogen atoms are forced (fused) together under a certain amount of *pressure* and *temperature* ($3.8x10^{12}$ psi pressure, causing 15 million °C / 27 million °F, heat to start a thermonuclear reaction) to form a heavier element, the *helium* atom.

Fusion takes place naturally in the core of all types / classes of stars, it will provide the stars with nuclear force energy to stay alive and display *heat* and *light*.

NUCLEAR FISSION

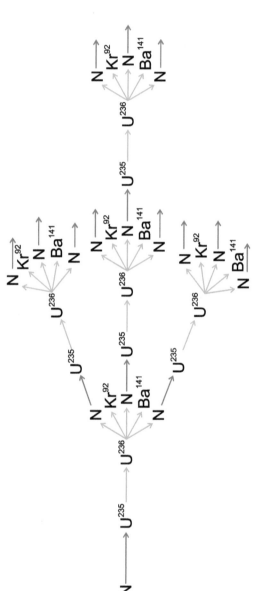

*Nuclear Fission is the fragmentation of a heavy atomic Nucleus into two roughly equal masses, plus three fast Neutrons and tremendous amount of nuclear Energy.
The three fast Neutrons must be slowed down by Neutron moderators into "Thermal Neutrons" in order to create a controlled nuclear chain reaction.

→ Chain Reaction

Fig. 21

NUCLEAR FISSION

It is a nuclear process exactly the opposite of *nuclear fusion*. In it, instead of two light atoms being fused together, a heavy atom such as uranium (U-92) or plutonium (Pu-94) is bombarded with *thermal neutrons*, producing a heavier and much less stable fissionable radioisotope. The new unstable radioisotope will instantly split into two lighter elements plus a huge amount of energy. For example, U-235 atoms are bombarded with *thermal neutrons,* producing U-236, which is a very unstable radioisotope, splitting or decaying into two lighter elements of: krypton (Kr-92) with mass number 92 and barium (Ba-141) with mass number 141, plus, three *free high-speed neutrons* (note: 141+92+3=236).

In turn, these three high-speed free neutrons will be slowed down by the help of the *moderator elements* inside the nuclear reactors to become *thermal neutrons*. Each of these thermal neutrons will impregnate another uranium 235 atom, starting a spontaneous *nuclear fission chain reaction*.

THE HISTORY OF NUCLEAR FISSION

A few years before World War II, the idea of nuclear fission was suggested by the Italian-American theoretical physicist *Enrico Fermi*. Then the idea was picked up by *Lise Meitner,* an Austrian-Swedish nuclear physicist who mathematically proved that *nuclear fission* would produce tremendous amounts of *energy*. Meitner shared her formula with her ex-boss, *Otto Hahn,* a German chemist. Later, *Hahn* and his partner, *Fritz Strassmann,* worked on the formula and came up with a successful experiment of the processes of *nuclear fission.*

In 1945, for the first time, two atomic bombs were used against Japan in order to end the war. Hahn and his partner, F. Strassmann, were awarded the Nobel Prize in Chemistry. In 1944, none of the nominees for Nobel Prize in Chemistry were qualified. Then, in 1945, it was given to Hahn and his partner for his discovery of the fission of heavy nuclei.

Unfortunately, Lise Meitner, the nuclear physicist who wrote the formula for nuclear fission, was not included in sharing the Nobel Prize along with Hahn and Strassmann. It was a shame, but years later, they named the element 109 in the periodic table of elements in her honor (Meitnerium-109 / Mt-109).

THE ATOMIC BOMB

In 1942, at Los Alamos, New Mexico, a group of physicists, chemists, metallurgists, explosive experts, and thousands of technicians were gathered to work under the scientific direction of *J Robert Oppenheimer* to build the first atom bomb.

Of course, it was all under the supervision of the Army with the code name *The Manhattan Project.*

The first atomic bomb, which was a *uranium-based* design, nicknamed "Little Boy," was dropped on Hiroshima on August 6, 1945, destroying an area of five square miles. Then, the second bomb, a *plutonium-based* design, nicknamed "Fat Man," was dropped on Nagasaki on August 9, 1945, destroying an area of three-square miles. And that made Japan surrender on August 14, 1945.

Fortunately, the world community learned from the

disastrous outcome of WWII, and they decided that a change of course was in order.

In 1964, President Lyndon B. Johnson allowed the use of nuclear fission technology, which was perfected by the Manhattan Project engineers, to be used by private ownership.

Since then, many humanitarian uses of this field have been up and going. These include the nuclear reactors generating electrical power and the production of radiopharmaceuticals used in diagnostic and therapeutic procedures in nuclear medicine, PET scans, radiation therapy, and many more. First, we must believe that because of the Manhattan Project, we have saved millions of lives by stopping the war, saving trillions of barrels of crude oil, billions of tons of coal, millions of acres of forests, much cleaner air, and many more benefits because of the nuclear technology. However, we must be more vigilant about the management of the atomic wastes and their byproducts!

Nuclear wastes are harmful to our environment and some of their byproducts will live for thousands of years to come!

THE ENRICHMENT OF U-235

When uranium ore is mined, it is a mixture of three uranium isotopes: uranium-238, which comprises 99.25% of the sample; uranium-235, comprising 0.72% of the sample; and uranium-234, with 0.003% of the sample.

The fact that only U-235 can be used in nuclear power plants and atom bombs, the uranium ore must be purified or enriched. There are two grades of enrichments. The

SCHEMATIC OF A NUCLEAR POWER PLANT

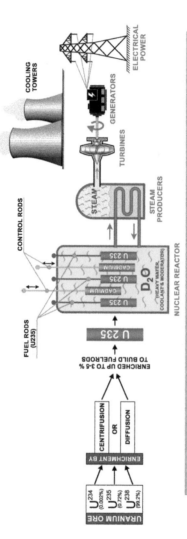

Control Rods: To control the fission rate, made of strong neutron absorbers such as: Cadmium, Boron, Silver or Indium rods which are used to either shield or expose the Fuel Rods (U235).

Moderators: Such as Heavy Water Deuterium D_2O (act as coolant & moderator) or pure Graphite, will slow down the fast Neutrons into "Thermal Neutrons" which are suitable to impregnate U-235 atoms into U-236 which is highly Fissionable.

REACTOR PRODUCES HEAT	HEAT PRODUCES STEAM	STEAM TURNS TURBINES	TURBINES ROTATE GENERATORS TO PRODUCE ELECTRICAL POWER

Nuclear Power Plants Disasters: It takes Place mainly Because of the malfunctions of the cooling system or control Rods. It happens in two stages:
First stage - Runaway;
Second stage - Meltdown.

Fig. 22

first type is called Low Enriched Uranium (LEU), which is used for peaceful purposes such as nuclear power plants (3% to 5% enriched). The second type is Highly Enriched Uranium (HEU), which is used in nuclear weaponry (20% to 90% enriched).

The enrichment processes of U-235 can be done by *physically* separating the U-235 from U-238. This physical enrichment is done in two ways:

1. *Centrifuging:* The sample is placed in a *centrifuge* with a supersonic speed, which will separate the two isotopes of U-238 and U-235 on the basis of their different atomic masses.

2. *Diffusion:* The sample is forced through a *delicate metallic membrane* in order to physically separate the two isotopes.

After the enrichment processes, the enriched U-235 (3% to 5% enriched) will be placed in a nuclear reactor power plant to produce electricity.

Fig. 22 shows a schematic of a power plant from the mining of the uranium ore to the production of the electricity, lighting up the cities.

NUCLEAR REACTOR MAIN PARTS

The atomic fission processes inside a nuclear reactor will provide *thermal power* to *heat* and *steam* the water. In turn, the *steam* will move the *turbines* and the *turbines* will move the *generators* that will generate *electrical power*.

In a nuclear power plant, there are four main elements including:

    1. fuel rods

2. moderators
3. control rods
4. coolant systems

The rest are auxiliary parts transforming the heat and steam into electrical power.

THE FUEL RODS

Fuel rods are made of *uranium* (U-235), which will be bombarded by *thermal neutrons* to be transformed into a highly fissionable uranium 236.

THE MODERATORS

Nuclear moderators, such as *pure graphite* or *heavy water,* can slow down *fast neutrons,* turning them into *thermal neutrons* in order to facilitate the processes of nuclear fission. Because the neutrons that are produced from the fission processes are too fast to be able to impregnate the U-235 in the fuel rods. The moderators are the kind of elements that do not absorb the neutrons; otherwise they would cause neutron depletion.

THE CONTROL RODS

The control rods are made of *cadmium, boron*, *silver*, or *indium* to either *shield* or *expose* the fuel rods (U-235) from the *thermal neutrons*, thus controlling the *chain reaction*. Control rods are strong neutron absorbers and their movements between the fuel rods either slow down or stop the nuclear chain reactions altogether (Fig. 22).

THE COOLANT SYSTEMS

Heavy water (deuterium / D_2O) is used as coolant, which is the most outer layer of the system, to keep the plant cool and prevent the nuclear power plant's runaway or meltdown. If by any reason the reactor is not properly maintained and watched, disaster will strike.

The first stage of disaster is called the *runaway,* and the second and the last stage is called the *meltdown!*

One bit of good news is that after the meltdown, a *nuclear explosion* would not happen. In the past, a few nuclear reactor accidents have happened. In 1979, at the *Three Mile Island* power plant in the US, a partial meltdown took place because of the cooling system's malfunction. In 1986, at the *Chernobyl* nuclear power plant in Ukraine, a complete meltdown took place because of a faulty design and human error.

Then, the last one took place in 2011: the *Fukushima Daiichi* nuclear power plan disaster, in Japan. The cause of the accident was a 9.0 magnitude earthquake, causing a 50-feet tsunami that disabled the power supply of the cooling system.

CHAPTER TEN

"All these fifty years of conscious brooding have brought me no nearer to the question of 'What are light quanta?' Every Tom, Dick and Harry thinks he knows it, but he is mistaken."
—Albert Einstein

THE NUCLEAR AGE

The nuclear industries and their products are so intertwined with our daily lives that it would be hard to imagine life without them.

It all started with the discovery of the *X-ray* in 1895 by Wilhelm Röntgen, a German physicist. Just one year later, in 1896, the *radioactivity* was discovered by Henri Becquerel. And again, a year later in 1897, the discovery of *electrons* by J.J. Thomson energized scientific societies throughout the world, and the race for discoveries of more elementary particles was on.

DEVELOPMENT OF THE ATOM BOMB

Discoveries of the subatomic particles, and later the elementary particles, led to the development of the theory of *the Standard Model of the elementary particles.* It was the beginning of a new era—the dawn of the atom bomb! The rise of the genie out of the oil lamp after 13.8 billion years!

After Lise Meitner, Otto Hahn, and Fritz Strassmann unlocked the mysteries of the hidden power within the uranium atom, the genie was out of the lamp and stood humbly to fulfill his master's wishes.

The genie received his orders from the masters and flew twice over the two beautiful, populated cities of Hiroshima and Nagasaki, pounding on his chest, huffing and puffing high winds, massive shocks, pressure waves, radiation burns, tearing and breaking, contaminating the soil and the water, leaving behind nothing but massive destruction, death, pain, blood, and tears!

"Your third and final wish, my masters?" the genie asked humbly.

The masters themselves were shocked and distraught, looking at each other thinking, "How are we going to put the genie back in the lamp?"

"No! We don't need to put him back in the lamp!" the masters said in their huddle. "We have better plans!"

"Genie!"

"Yes, masters!"

"We want you to light up all our cities around the world; we want you to help our industries to be more efficient; we want you to get involved in our medical field, healing our sick; we want you to help us to explore the new frontiers in the universe . . ."

"B-but, my masters! You have only one more wish left to ask of me!" the genie said with a groan.

"Maybe he wants to go back in the lamp?" one of the masters said carelessly.

A moment of eerie silence went by.

"Okay . . ." the genie said sheepishly.

And the genie was hired.

THE UNITED NATIONS CHARTER

In 1945, right after WWII, representatives of fifty countries met in San Francisco and drew up *the United Nations charter* to uphold the human rights of citizens of the world—a charter that could help them to achieve a higher standard of living and address their social, economic and health problems. The charter also stressed freedom for all citizens without distinction as to race, sex, language, or religion.

UN charters' article 103 states that obligations to the UN charters prevail over all other treaties.

The destructive use of nuclear force at the end of the World War II and the realization of its destructive magnitude brought a new sense of urgency to all members of the world community to see themselves as a whole and not separate in any way, shape, or color.

There is a poem etched at the entrance of the UN, written by Saadi Shirazi, a thirteenth century Persian poet, that declares:

All beings are members of a whole

In creation of one essence and soul
If one member is afflicted with pain
Other members uneasy will remain
If you have no sympathy for human pain
The name of human you cannot retain.

Wow! So simply said yet so powerful to bring together the whole humanity as one.

Finally, we all came to our senses and the *UN's charter* helped billions of people in hundreds of countries to get out of the ditches of future wars, poverty, prejudice, exclusion, and indifference.

Now, the same technology of nuclear force is being used to save our world from depletion of its resources by using the energy from the nuclear power plants. Also, nuclear technology saves millions of lives every year in medical fields by providing millions of different types of radiopharmaceuticals.

There are many more benefits from nuclear technology in all walks of life: building *atomic batteries* to power spacecrafts and satellites, helping us in our space endeavors, airport securities, and the steel industries for quality controls. In agriculture, radioisotopes are being used to develop new strains of agricultural crops that are drought and disease resistant. Radioactivity is also used for insect and pest management, soil quality control, and much more.

One of my intentions with this book is to explain the variety of ways that atomic industries have been instrumental in medical fields and healthcare.

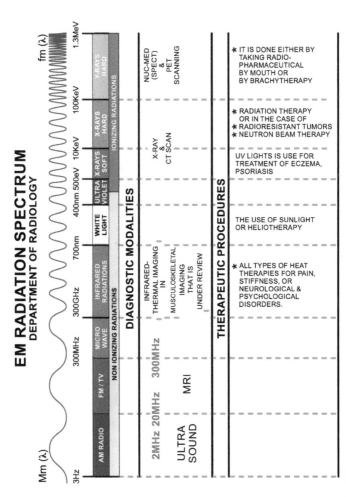

Fig. 23

THE USE OF EM RADIATION IN MEDICAL FIELDS

To start learning about the use of EM radiation in the Department of Radiology as a whole, it will be helpful in distinguishing between the two categories of medical fields in the Department of Radiology.

Within the Department of Radiology, a number of departments are using *non-ionizing* radiation, and the rest are using *ionizing* radiation. Non-ionizing radiation is weaker and unharmful, but ionizing radiation is more energetic and able to ionize / damage any atom, including human cells.

THE DEPARTMENT OF RADIOLOGY

The history of the medical field started with quackery, witch doctors, herbal medicine, and many more of that nature. Finally, in late tenth century to early eleventh century, some pioneer physicians such as Mohammad Zakariya Razi and Abu Ali Sina, two Iranian physicians among other European thinkers, brightened the Medieval European medical paths by writing the *medical textbooks* in surgery and therapy that are still being taught in universities around the world.

Zakariya Razi has been described as the father of *pediatrics*, and a pioneer of *obstetrics* and *ophthalmology*.

Although there have been many advances in the medical field, the biggest breakthrough came about 127 years ago, by the discovery of the *X-ray* in 1895 and *radioactivity* in 1896, which revolutionized the world of medicine by opening a new floodgate to the *noninvasive* medical *diagnostic* and *therapeutic* procedures.

Nowadays, not only X-rays and gamma radiations, but the whole *spectrum* of EM radiation is being utilized in all facets of the medical field.

Before I learned about EM radiation, I never thought in my wildest imagination that all the machineries in different sub-departments of the Department of Radiology generate and detect the same EM radiation (photons). The only difference between all these modalities is the usage of different *frequencies* of radiations being produced by their machineries.

Some machines generate and use *non-ionizing radiations*, such as ultrasound (US) and Magnetic Resonance Imaging (MRI), as opposed to X-ray machines, which generate *ionizing X-rays,* and in the Department of Nuclear Medicine and Positron Emission Tomography, gamma cameras are equipped to detect *ionizing gamma radiations* (Fig. 23).

USE OF NON-IONIZING RADIATION

Modalities using non-ionizing radiation include *ultrasound* (US), *Magnetic Resonance Imaging* (MRI), and *microwave* radiation. For example, microwave radiation is used for treatments of cancerous tumors, warts, and will stop bleeding during surgery. *Infrared* radiation therapy is being used for pain, fatigue, stiffness, and the treatment of neurological and psychological disorders.

There are other non-ionizing radiation therapies such as *heliotherapy,* or phototherapy, which contains non-ionizing UV lights (the first half of the UV light band) from the sun, that kills all kinds of bacteria. It also cures some skin disorders, such as *eczema* and *psoriasis*.

USE OF IONIZING RADIATION

The *ionizing radiation* is a totally different story!

I have a simple analogy that I hope will help you to differentiate between ionizing and non-ionizing radiation.

Imagine there are two vehicles rolling down a hill at 100 miles per hour. One of the vehicles is a loaded eighteen-wheeler, and the other vehicle is a Volkswagen Beetle (note: all EM radiation, regardless of frequency or strength, travels with the same speed of light).

If both vehicles are out of control, and both crash into a concrete wall at the bottom of the hill, which vehicle do you think will go through the concrete wall and cause more damage beyond the wall?

Ionizing radiation would be the loaded eighteen-wheeler and the Volkswagen Beetle would be the non-ionizing radiation.

That is why so much attention is being paid to handling, transporting, and using ionizing radiation and radioactive materials as opposed to the usage of non-ionizing radiation. That's why in the hospitals, doctors, nurses, and technologists wear a heavy lead apron to protect their body parts against the ionizing radiation that they happen to be working with. Therefore, you should avoid places where the radioactive signs are displayed. If you do go, protect yourself and follow the experts' directions.

However, as much as *ionizing radiation* is destructive and scary, when it is used *properly* and *purposefully*, it is a blessing in disguise!

All the diagnostic and therapeutic procedures that are done in the Department of Radiology are *noninvasive*; the amounts of radiopharmaceuticals or X-rays that are being used are safe; and they are heavily regulated by a super knowledgeable group of professionals in the country known as the Nuclear Regulatory Commission (NRC).

What comes after the diagnostic procedures would be a wide range of therapeutic options, and the doctor will discuss them with his patients. The options can be done singly or with a combination of a few choices, such as medication, radiation therapy, surgery, or a noninvasive therapy through the usage on ionizing radiopharmaceuticals.

CHAPTER ELEVEN

"A physicist is just an atom's way of looking at itself."
—Niels Bohr

DIAGNOSTIC AND THERAPEUTIC PROCEDURES

There are many modalities of performing diagnostic and therapeutic procedures. However, there are two main categories of non-ionizing and ionizing machineries, each with their own subspecialized sub-departments:

I. Modalities using non-ionizing radiation
   a. Ultrasound / Ultrasonography (US)
   b. Magnetic Resonance Imaging (MRI)

II. Modalities using ionizing radiation
   a. X-ray and CT
   b. Nuclear Medicine and PET scan

MODALITIES USING NON-IONIZING RADIATION

ULTRASOUND / ULTRASONOGRAPHY

Ultrasound is a noninvasive diagnostic medical imaging using the same technology as radar. Ultrasound machines send a sound pulse with a frequency between 2 MHz to 20 MHz. However, one must keep in mind that the lower frequencies (longer wavelength) will penetrate deeper into the body than the higher frequencies (shorter wavelength). Although the lower frequencies travel deeper, they produce a *lower image quality,* whereas the higher frequencies penetrate less, producing better *quality images*.

Note: An example of this phenomenon takes place in our daily lives. When you are walking on a beach and being mesmerized by the sunset as the sun disappears over the horizon, those beautiful red and orange lights that you see are the *refracted* lower frequencies—a longer wavelength portion of the sun's white lights.

It is because the high frequencies (short wavelength) of the white light have been weakened and eventually stopped by the atmospheric particles, while the lower frequencies (long wavelength) are going through the atmosphere with not much interaction. Red and orange are the lowest frequencies and the longest wavelength of the white light.

In ultrasound imaging, when sound pulses are sent into the body using a *probe* or *transducer*, the soundwaves travel into the body until it hits a boundary between two different tissues, such as soft tissue and bone.

The denser the tissue, the more soundwaves will bounce or echo back to the transducer. Then, the signal from the transducer is transmitted to the machine, where the *time* between the soundwaves transmitted to the tissue / organ and the return time (from and to the probe), and the *intensity* of the returned signals are calculated, and an image is formed.

Ultrasound frequencies are non-ionizing radiation located at the beginning of the EM spectrum, where the RF are. They have the *lowest frequencies* with the lowest *energies*.

Ultrasound procedures are among the safest diagnostic procedures in hospitals.

Obstetric ultrasonography is one procedure that is a part of *prenatal* care done on a regular basis for monitoring the progress of a fetus as early as five weeks.

Doppler exam is a routine procedure done for the circulatory problems of the legs.

Echocardiogram (or *echo cardio*) is another fast and easy noninvasive procedures done on patients with cardiac problems.

MAGNETIC RESONANCE IMAGING (MRI)

The MRI is also a noninvasive diagnostic medical imaging that uses a magnetic field (MF) as a tool to reveal the inner world of the human body.

MRI provides doctors with high-resolution images using harmless RF rather than ionizing radiation. In some cases, the use of *intravenous contrasts* will enhance the image quality of the two adjacent groups of tissues with very similar densities.

Unlike the ultrasound machines that are the size of a

MR I

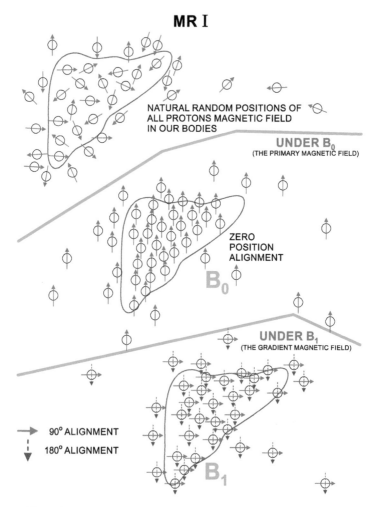

NATURAL RANDOM POSITIONS OF
ALL PROTONS MAGNETIC FIELD
IN OUR BODIES

UNDER B$_0$
(THE PRIMARY MAGNETIC FIELD)

ZERO
POSITION
ALIGNMENT

B$_0$

UNDER B$_1$
(THE GRADIENT MAGNETIC FIELD)

B$_1$

→ 90° ALIGNMENT
⇣ 180° ALIGNMENT

Fig. 24

cash-register, the MRI machines are the size of a small pickup truck in order to accommodate for whole-body imaging.

The main idea of an MRI procedure is to *manipulate* the *magnetic field* of a living body with a much stronger *magnetic field* created by the MRI machines.

First, let's see what constitutes the MF within the human body.

In physics, it is a fact that any piece of matter, regardless of its size—whether it is as massive as a star or as little as a subatomic particle—if it *spins*, it can create its own magnetic field.

For example, in the Standard Model of the elementary particles, protons are part of fermions, and all fermions possess one half-spin, therefore, all protons possess one half-spin, and therefore, all protons create their own magnetic fields around them.

Now, let's assess the proton situation in human bodies.

Water constitutes about 65% of our bodies, and each water molecule (H_2O) contains 10 protons (P) in its make up. Therefore there are trillions and trillions of protons in our bodies, each with their own magnetic field around them. One might ask, if there is such a huge number of magnetic fields in our bodies, why are our bodies not walking magnets?

The reason is because all the protons in our bodies are floating randomly in different directions. However, if a human body gets in a situation where there is a strong magnetic field around it, the strong magnetic field will align all the protons in the body in one direction, turning the human body into an organic magnet! Again, one might ask why the Earth's magnetic field can't do that to our bodies. The answer is, the Earth's MF is very weak, around 0.00005 Tesla. So, it has no effects on our bodies and we are okay!

MRI procedures are the modality of choice for *soft tissue imaging* such as: brain-aneurysms, tumors, spinal cords, nerves, muscles, ligaments, and tendons. Furthermore, MRI procedures are preferred over CT imaging, because CT procedures are done using ionizing radiation.

Now let's see how MRI machines use the randomness of the proton's spin-direction in our bodies to its advantage.

MRI MACHINES

MRI machines are made of four distinct parts.
- The Primary Magnetic Field (Bø)
- The Gradient Magnetic Field (B1)—the second magnetic field
- The RF Coils
- The Computer System

THE PRIMARY MAGNETIC FIELD (Bø)

When a patient is placed inside a tube-like structure in an MRI machine, the machine's Primary MF (Bø), with the strength of 1.5 Tesla in old machines and 3 Tesla in new machines (note: a chunk of magnet in a junkyard with a strength of 0.3 Tesla can lift a pick-up truck) will align all the protons in the patient's body into one direction of *zero position* (contrary to natural random positions of all protons in our bodies).

At this point, the patient's whole body is technically

turned into a magnet.

THE GRADIENT MAGNETIC FIELD (B1)

The gradient MF (M1) will override the Primary MF (Bø) and align all protons into either a 90-degree or 180-degree position, from the zero position of Bø (Fig. 24). During these processes where protons assume the second position, the protons will *absorb* some *energy* in order to override the Primary position of MF (Bø) and stay at 90- or 180-degree positions from the zero-degree positions.

RF COILS / PULSES

The radio frequency pulses will act as a switch, turning the gradient MF (B1) on or off.

When the *RF signals* turn *on / activate* the gradient MF, all the protons will *absorb* some *energy* and move into a new position of either 90 or 180 degrees from the initial Bθ position. Then, when the RF *signals* turn *off* the B1 gradient MF, all the protons, by *emitting / losing* the *energy* that they had gained earlier, will fall back into the Primary Bθ position of zero.

Now, this turning *on* and *off* of the gradient MF (B1) is the bread and butter of the MRI machines' imaging.

When all the protons are in the position of 90 or 180 degree and the gradient MF (B1) is turned off, the amount of *time* that it takes for the protons to fall back into the position zero of Bø, and the *amount of energy* that is *emitted* by protons while falling back, are *unique* to every single proton / tissue and it is called the *signature signal.* These *signature signals* will differentiate between different tissues in the body.

These *signature signals* not only differentiate between different tissues, but they also differentiate between healthy tissues and altered tissues, such as tumors and torn ligaments.

THE COMPUTER SYSTEM

The computer system will analyze these *signature signals* and distinguish between different tissues or even the smallest change within the same tissue.

The first clinical MRI images were obtained in 1977, displaying another monumental achievement in noninvasive diagnostic procedures in the medical field.

CHAPTER TWELVE

"If you quit once, it becomes a habit. Never quit."
—Michael Jordan

MODALITIES USING IONIZING RADIATION

X-RAY AND CT

In 1895, the German mechanical engineer and physicist, *Wilhelm Conrad Röntgen*, discovered and produced an EM radiation in the wavelength-band known as X-rays. Röntgen, for this achievement, was awarded a Nobel Prize in physics in 1901. He was also honored by having his name given to element 111 in the periodic table of the elements as *roentgenium* (Rg 111).

X-RAY MACHINE PARTS

X-ray machine is composed of six main parts.

X - RAY TUBE

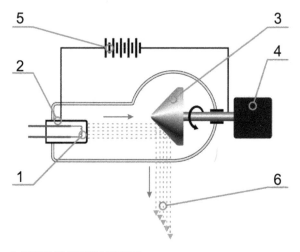

1 - HEATER: TO HEAT THE CATHODE.
2 - CATHODE: HEATED CATHODE WILL RELEASE ELECTRONS (ē).
3 - ANODE: WILL ATTRACT THE ELECTRONS AND REFLECT THEM OUT.
4 - MOTOR: WILL ROTATE THE ANODE TO PREVENT BURN.
5 - DC POWER SUPPLY: TO CREATE ELECTRICAL POTENTIAL BETWEEN
     CATHODE & ANODE.
6 - X-RAY BEAM: AIMING AT THE PATIENT.

Fig. 25

Vacuum tube

Housing inside of it the two electrodes of *cathode* and *anode*.

Cathode

The cathode is the electrode made of *tungsten or aluminum filament* that, when it is electrically heated into a glowing temperature, releases *high-speed electrons.* These electrons will be focused into a beam of electrons heading toward the *anode plate*.

Anode

Anode is a positively charged electrode made of *tungsten plate*, which attracts the high-speed beam of negatively charged electrons emitted from the cathode. When the beam of electrons hits the anode plate, it will produce 90% heat and 1% X-rays.

Motor

For rotating the *tungsten plate (anode)*, in order to prevent it from melting by the high-speed electrons.

Coolant

To keep the system and the vacuum tube cool.

Lead shield

In order to protect people around the X-ray machine when it is "on."

HOW AN X-RAY MACHINE WORKS

In an X-ray machine, when the cathode is heated, it will produce a stream of very high-speed *electrons* in the vacuum space of the tube.

The electrons, having negative charges, are attracted to the positively charged *anode* and hit the tungsten plate with force. The collisions between the high-speed electrons and the tungsten atoms will produce either *characteristic X-rays* or *Bremsstrahlung radiation*, which are commonly known as X-rays.

Then, the X-ray will go through a filter and a collimator to get adjusted for the area on a patient's body to be imaged

To take an X-ray image, the patient is either standing or lying down on a table, with an unexposed film behind them. As the X-ray goes through the patient's body, it will be absorbed differently by different tissues in the body and will expose the photographic film behind them with different intensity in which an image will be produced.

Since X-ray imaging cannot clearly distinguish between the two adjacent soft tissues, in some cases, the use of a *contrast* is necessary.

For example, in *angiography*, to visualize the blood vessels embedded in soft tissues, the administration of an intravenous contrast will be necessary for visualization of the whole tree of blood vessels.

The contrast media will absorb X-rays more effectively than the surrounding tissue, providing a better contrast between the two different soft tissues.

Or in the case of digestive system, a contrast will be given orally for a better visualization of the patient's digestive tract.

FLUOROSCOPE AND CONTRASTS

Contrast media is often used in conjunction with a *fluoroscopy*. It allows the interventional radiologist to perform a diagnostic, therapeutic, or even surgical procedure on a patient's internal organ. The radiologist will be watching the real-time videos of the procedure on a monitor.

There are many procedures done by the use of fluoroscopes and contrasts, such as angiography, angioplasty, stent-placements, tumor ablations, needle biopsies, and many more.

There are two other modalities that use *hard X-rays* (10 KeV to 100 KeV).

One is *mammography*, which is a noninvasive diagnostic imaging of the breast done on male and female patients. However, the male mammography comprises less than 1% of all mammograms being done.

Unlike the regular X-ray imaging, in mammography, the breast tissue is placed between two plates and squeezed rather firmly for a better resolution.

The other modality using *hard X-rays* is the *computed tomography (CT),* or *Computerized Axial Tomography (CAT)* scanners. The CAT scanners are bigger and more sophisticated than X-ray machines for the purposes of providing whole-body, three-dimensional images.

Nowadays, there are *multimodality scanners* performing two types of imaging at the same time. Scanners such as CT-MRI, CT-PET, or SPECT-CT, which will save time and money. Usually, in performing a multimodality procedure, it will be a combination of a *functional* modality such as nuclear medicine SPECT or PET scanners, with an *anatomical* modality such as CT or MRI.

Another advantage of using a multimodality procedure is for a better *pinpointing* of the affected area to help the surgeons.

THERAPEUTIC USE OF X-RAYS: RADIATION THERAPY (RT)

A high-energy X-ray or electron beam produced by a *Linear Accelerator* (LINAC) will be aimed at a patient's specific organ, for the purpose of controlling or destroying the unwanted tissues or tumors within that organ. And the goal is to not damage the surrounding tissues.

Radiation therapy takes advantage of the fact that the cells that are being divided rapidly (cancerous cells) and are thus particularly more sensitive to being damaged by radiation than the healthy, older tissues.

However, in some cases, there are certain types of cancerous cells / tumors that are resistant to *X-ray radiation*; therefore, the oncologist will resort to *hadron* therapy, which involves either *neutron beam therapy* or *proton beam therapy*.

Furthermore, in some cases, a radiation source is implanted inside a patient's body, near or inside the target organ, to destroy or control the growth of unwanted tissues. This procedure is called *Brachytherapy*.

CHAPTER THIRTEEN

"Winning isn't everything but wanting to win is."
—Vince Lombardi

NUCLEAR MEDICINE AND PET SCAN

X-rays and radioactive materials are both ionizing radiations; however, *X-rays*, are dry radiation with less ionizing powers than the radioactive materials being used in nuclear medicine and PET scanning. Radioactive materials are more ionizing and can be found in three forms of gases and solids, but mostly in liquid forms.

Before continuing, I would like to make something clear, and that is the concept of *protecting ourselves* from the harmful effects of ionizing radiation.

When you are in an X-ray department, you protect yourself from active *X-ray machines,* which lasts only for a few seconds. But when you are in the Department of Nuclear Medicine or PET scanning, you must protect yourself from *hot patients.* By saying "hot," I don't mean beautiful or sexy, but a patient who has been injected with a radiopharmaceutical for a procedure.

Unlike the X-ray machines that are active only for a few seconds, a hot patient must be avoided for a few hours up to few days, depending on the type of the radiopharmaceutical that has been administered to them.

Just remember that the *X-ray machines* produce X-rays, and they are harmful only when they are *shooting X-rays*. And when you are in a department of nuclear medicine, remember that a *gamma camera* itself is as harmless as a handheld photographic camera; the only difference is that the regular cameras are sensitive to visible light while gamma cameras are sensitive to invisible gamma radiation.

When a nuclear medicine technologist administers a radioactive material (radiopharmaceutical) to a patient as a *gas* to breathe in, or as a *capsule* to swallow, or as an intravenous injection liquid, then the patients themselves become a *source* of gamma radiation and we must protect ourselves from *them.*

So, next time you are around an X-ray machine that is about to be turned *on,* protect yourself by putting an effective distance between yourself and the machine. And if you are near a *gamma camera,* being *on* or *off,* don't be concerned or alarmed at all. However, if you are around a *hot* patient and you need to be near them, ask the technologist how you can protect yourself. The NRC recommends standing at a distance of six feet or more, spending as little time as possible near the patient, and keeping a good shielding between you and them.

At this point, I believe a short history of discoveries of radioactivity and their applications in the medical field will be helpful.

Discovery of the X-ray in 1895 by Wilhelm Conrad Röntgen was the spark that enlightened the world of noninvasive medical procedures; and the discovery of *radioactivity* by the French physicist, Henry Becquerel in 1896, was the renaissance of the subatomic discoveries. Together, they opened the door to noninvasive diagnostic and therapeutic procedures in the medical profession.

A SHORT OVERVIEW OF NUCLEAR MEDICINE HISTORY

⇒   1895, the discovery of X-ray by Wilhelm Conrad Rontgen, a German mechanical engineer and physicist.

⇒   In 1896, the radioactivity emitted from the element uranium (U-92) was discovered by Henry Becquerel, the French physicist. However, nothing was known about alpha or beta particles or gamma radiations.

⇒   In 1898, the discovery of polonium and radium by Marie Curie, a Polish-born and naturalized French physicist and chemist, and her husband, Pierre Curie, the French physicist. They performed the first bone scan ever using radium. Their technique was called shadow imaging, because a patient was placed between the radium source and a photographic plate. They also together coined the name "Radioactivity." Still, nothing was known about alpha, beta, or gamma radiation!

⇒   1898, was the year of the discovery of alpha (α) and beta (β) particles emitted from the radioactive element of uranium by Ernest Rutherford. He was one of

J.J. Thomson's students. (J.J. Thompson discovered the electron in 1897.)

⇒ 1900, two years after the discoveries of the alpha (α) and beta (β) particles, the French chemist Paul Villard discovered gamma radiation (γ) while working on the radioactive element of radium (Ra-88).

⇒ 1911, the cloud chamber or Wilson cloud chamber was invented by Charles Wilson, a Scottish physicist and meteorologist. It is a particle detector used for visualizing the passage of ionizing particles through a sealed environment of water or alcohol vapor. The ionizing particles leave a characteristic trail of condensation in the vapor.

⇒ 1927, for the first time, radioactive radon (Rn-86) was injected into a patient's arm to measure the speed of blood flow from one arm to the other arm. In this experiment, a cloud chamber was used as the radiation detector.

⇒ 1928, Hans Geiger and Walther Muller, German physicists, developed the Geiger counter. It became a practical instrument for detection and measurement of the ionizing particles such as alpha (α), beta (β), and gamma radiation (γ). The Geiger counter is a very useful instrument in nuclear medicine. Earlier, in 1911, Geiger had helped Rutherford with his gold foil experiment as well.

⇒ 1930, the cyclotron (a particle accelerator) was invented by Ernest Lawrence. He was an American nuclear scientist at UC Berkeley. A cyclotron produces the synthetic radioactive elements now being used in the Department of Nuclear Medicine. He also worked on enrichment of uranium during the Manhattan Project. After WWII, the Manhattan Project's atomic reactors were put to work for the production of medical radioactive materials.

⇒   1937, Carlo Perrier, an Italian mineralogist, and Emilio G. Segre, an Italian American physicist, discovered and produced the element technetium (Tc-43) synthetically. Technetium was one of the four elements that was predicted by Dmitri Mendeleev in 1869, in his periodic table of the elements.

⇒   1951, the first rectilinear scanner was built by Benedict Cassen. It was a defining invention in the evolution of clinical nuclear medicine's imaging. The rectilinear scanner was comprised of a motorized scintillation detector coupled to a relay printer. The first scan was done on a thyroid gland after the patient swallowed a radioactive iodine capsule. A small detector comprised of one small scintillation detector would be placed a few inches from a patient's thyroid gland and scanned from right to left and from top to bottom, stopping 15–20 seconds at each point. Then, a pixelated image of thyroid gland would be printed on a piece of paper across the room. The image looked like a bunch of dots displaying an image with the worst resolution. But it was a huge start for nuclear medicine imaging.

⇒   1951, Hal Oscar Anger, an American electrical engineer and biophysicist at UC Berkeley, invented a device for the measuring of radioactivity in a small sample of urine or blood, and he called it the well counter. It is made of a small sodium iodide crystal detector on top of a photo multiplier tube (PM tube) inside a protective lead tube. The well counter is a smaller version of a gamma camera before the actual gamma camera was invented by the same man. Later, in 1958, he developed the scintillation camera or gamma camera, which is the same concept as the well counter. A scintillation camera or gamma camera is made of one big plate of sodium iodide crystal detector, and instead of one photomultiplier (PM tube) being placed behind it, 50–60 PM tubes are placed behind it to receive and amplify the weak signals coming from the crystal detector.

⇒   1958, the first Anger camera (the mother of all gamma cameras) was built by Hal Oscar Anger (he also built the well counter in 1951, the same year that Benedict Cassen built his rectilinear scanner), the American electrical engineer and biophysicist at UC Berkeley. His invention made the rectilinear scanners obsolete. Today's Anger cameras are called gamma cameras with the same concepts but much faster, more efficient, more versatile, and much better resolution!

⇒   1964, the production of the radioisotope Tc-99m in a cyclotron (cyclotron was invented by Ernest Lawrence in 1930) revolutionized the nuclear medicine procedures, in terms of its efficiency in having the right strength (eV), its perfect physical half-life (Tp), and for not overexposing the patients to unnecessary amount of radiation. I believe that the radioisotope Tc-99m to nuclear medicine is like the element titanium to racing bicycles.

⇒   1971, the first brain scan was performed using Tc 99m, for detection of glioma (brain tumor).

At this point, I believe it is better to explain the different aspects of nuclear medicine separately in order to make

them easier to follow.
- Nuclear medicine instruments
- Radiopharmaceuticals used in nuclear medicine
- Procedures
- Patient care

NUCLEAR MEDICINE INSTRUMENTS

There are three main pieces of equipment in the nuclear medicine field.

    1. Survey Meters

        a. Geiger counters

        b. Ion chambers

        c. Scintillation counters

    2. Radioisotope Dose Calibrators

    3. Gamma Cameras

SURVEY METERS

Survey meters are handheld ionizing radiation measurement instruments used to check for radioactive contamination on personnel, equipment, and the workplace. There are three types of survey meters:

Geiger Counters

It is a handheld ionizing radiation detection instrument, detecting and measuring alpha particles (α), beta particles (β), and gamma radiation (Υ).

Each Geiger counter is made of two parts: a sealed tube / chamber filled with gas and attached by a cable to the information box, which is the size of a toddler's shoebox. When ionizing radiation collides with the gas inside the sealed tube/chamber, some atoms inside the gas become ionized and an *ion pair* is created. For each collision, there will be a clicking sound and the meter will show the number of counts or collisions per minute (CPM).

This is the first instrument that every technologist will use to start their daily tasks. There are companies that deliver pre-calculated radiopharmaceuticals for each scheduled patient every day.

The technologist's responsibility is to use the Geiger counter to make sure that there is no leakage, contamination, or damages to the delivered boxes. Of course, all the readings must be recorded in a daily radiopharmaceutical logbook.

Ion Chambers

This one is also a handheld gas-filled ionizing radiation detector (argon gas), with no tube / chamber attachments. It measures the beta particles (β), gamma radiation (γ), and X-ray radiation.

However, it is the preferred means of measuring *high levels* of gamma radiation.

Scintillation Counters

This instrument is also for detection and measurement of ionizing radiation. It is made of scintillating material (sodium iodide [NaI], doped with thallium [Tl]). An incident particle such as electron, neutron, and alpha particles, X-rays, or high-energy photons will excite the material to emit / release visible photons. These photons are then converted into electrical pulses to measure the intensity and the energy of incident radiation.

All these different types of instruments for detecting and measuring of ionizing radiation show some level of contamination, but they are unable to differentiate between the kinds of radiation unless the operator has a good understanding of different types of ionizing radiation.

To determine different types of particles, one must implement a little personal *technique* and *interpretation* in obtaining a correct reading.

As we know, the alpha (α) particles only travel for 4 centimeters (1.6 inches) and can be stopped by a sheet of paper. And beta (β) particles travel for 4 meters and will be stopped by a sheet of aluminum.

The most energetic of all radiation, gamma radiation, can travel up to 400 meters, and a concrete wall or a thick sheet of lead is needed to stop it.

Here comes the technique that can help us in distinguishing different types of ionizing radiation in a contaminated area. It is known that an ion chamber does not detect alpha particles, and beta particles can only be effective within 4 meters (approximately 4 yards), and gamma travels much further. Therefore, utilizing the different particles' effective range and the type of instruments used can help us to get the correct measurements.

THE RADIOISOTOPE DOSE CALIBRATORS

The second most important piece of machinery in nuclear medicine is the radioisotope *dose calibrator* for measuring every single dose before being administered to a patient. It contains a cylindrical ionizing chamber that contains argon gas under high pressure, and it is airtight. Within the chamber, there are two electrodes with electrical potential between them. When a vial or a syringe containing the radiopharmaceutical is placed inside the chamber, the argon gas is ionized, producing an *ion pair* (ions of opposite electrical charges) migrating toward the anode and cathode, producing an electrical current flow between them. This current is proportional to the activities of the radiopharmaceutical being measured.

The *magnitude* of activities and the *type of radioisotopes* that is being calibrated can be selected manually on the calibrator's display unit. For example, the magnitude or amount of radiopharmaceutical being measured has a range within Curie or micro-Curie, and the type of radiopharmaceutical being measured can be selected as gallium, thallium, indium, iodine, or technetium.

According to NRC regulations, the administrated dose to a patient must be within ±20% of the prescribed dose (The 20% Rule). Furthermore, the dose calibrator must be capable of calibrating any dose as small as 10 µCi, as well as a multi-Curie dose.

GAMMA CAMERA

Gamma camera (γ-camera), also known as scintillation camera or Anger camera (invented by Hal Oscar Anger in 1958), is a device that constructs an image from the gamma radiation being emitted from a patient's body. This technique is known as *scintigraphy*.

A gamma camera is made of three main parts:

1. The Table

The table is the part that moves the patient in and out of the gantry, for the purpose of taking an all-around, three-dimensional, whole-body scan.

2. The Gantry and Camera Head

A *gantry* is a mechanical arm that holds the *camera head,* and it can maneuver it 360 degrees around the patient for imaging and constructing a three-dimensional image of the target organ or the whole body.

The *camera head* itself consists of a few main parts. The most outer part of the camera head facing the patient is called the *collimator*. It is a lens-like apparatus for channeling and directing the relevant beams of radiations (the radiations that are coming from the patient's body in a straight line and not in an angle), onto the *scintillation crystal* for capturing images with higher resolution, and at the same time the collimator absorbs the scattered or unwanted radiations.

Right behind the collimator is the *scintillation crystal*, made of *sodium iodide* that has the ability of stopping the gamma rays, and at the point of interaction, it will produce a flash of light.

Right behind the scintillation crystal, there are 50 to 60 photo multiplier tubes (PMT) amplifying the weak flashes of light, converting them into electrical signals, and sending them to the *computer* to be analyzed and constructed into an image.

It is also important to know that there are three types of collimators in terms of handling different radioisotopes with different energies (eV). They are called low-energy, medium-energy, and high-energy collimators.

Before any procedures, the technologist will change the collimator on the camera head according to the type of radioisotope being used.

There are also different shapes of collimators as well. Just like a regular handheld camera has different lenses for different tasks, the collimators have different designs for different sizes and views for various organs. There are four different designs of collimators: parallel hole collimators, pinhole collimators, converging collimators, and diverging collimators.

In general, think of a collimator as a honeycomb structure, covered by a protective, touch-sensitive plastic. To experiment and see how a collimator works, we can remove the plastic covering on the collimator, place a large family photo underneath it, and look through the holes: you would be able to see only a small portion of the picture from each hole. And that is the way a scintillation crystal detector will see a target organ in each procedure. If you take an image of an organ without a collimator, you would not be able to see anything but a big blob of dark cloud.

Since gamma radiation travels in every direction, a collimator will channel the influx of the radiation, accepting the ones coming straight from the patient and rejecting the ones that come at an angle.

Furthermore, collimators (those honeycombs structures) with smaller holes, taller bores, and thicker septal walls will produce sharper images.

The structure right behind the collimator is called a *scintillation crystal* (40 cm x 50 cm x 1 cm), which is typically made of *sodium iodide* single crystal with some mixture of thallium added (doped) for better activity. Thallium in this case is called the *scintillation activator*. Sodium iodide crystals give a tiny flash of light when gamma radiation (photons) hit it. In the production of *sodium iodide crystals,* three qualities are essential.

First, the higher the *density* (high Z#) of the crystal, the higher the probabilities of gamma rays' interactions with the crystal's atom. Some crystals are packed with diamonds for higher probabilities of interactions.

Second, the *brighter* the crystal the more visibility of light to photo multiplier tubes.

Third, the *faster* the better, in terms of interaction activities, because the visible pulses are in nanoseconds and more of the interactions will be recorded.

The last major part of the camera head is a set of 50 to 60 PMT, attached to the back of the scintillation crystal.

The flashes of light that have been created by the crystal are seen by these PM tubes and will be amplified and converted into an *electrical signal* and sent into the computer for analysis and construction of an image.

In short, when a *hot patient* (who has been administered a radiopharmaceutical) lays down under the camera head, the *collimator* will properly channel the gamma radiation onto the *crystal*; when the crystal is hit with the gamma rays it will turn the gamma radiation into a *speck of light* that will be seen by *PM tubes*, which is amplified and turned into *electrical signals,* and finally sent to the computer.

3. The Computer System

The *computer* will manipulate and convert the electrical signals into high-resolution, sharp *images* (Fig. 26).

BASIC PRINCIPLES
OF A GAMMA-CAMERAS

CAMERA HEAD

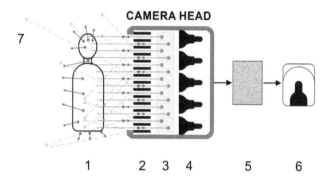

1 - Patient who has been administered with radiopharmaceutical.

2 - Collimator, allowing only the straight relevant beams of radiations going through.

3 - Scintillation detector NaI,TI crystal.

4 - Photo-multiplier tube array.

5 - Computer system & analog-to-digital converter (ADC).

6 - Monitor & Imaging

7 - Scattered, unwanted radiations.

●———●   Beams of radiations emanating from a patient who is injected with
a radiopharmaceutical.

●– – –▸   Beams of radiations emanating from a patient that has been reflected
from the plates of the callimator.

Fig. 26

CHAPTER FOURTEEN

"The genius of Einstein leads to Hiroshima."
—Pablo Picasso

". . . And then, to light up the cities around the world,
protecting the environment, and curing the sick."
—The Author

RADIOPHARMACEUTICALS

There are 118 elements in the periodic table of elements; however, this is only the tip of the iceberg. There are somewhere around 3,800 radioisotopes in nature. Only 200 of those are being used on a daily basis in different industries. And from that 200, only eight are being utilized in the field of nuclear medicine and PET scanning.

Out of these eight radioisotopes, some of them, such as I-123, I-131, Tl-201, and Xe-133 gas, can be administered to a patient without *tagging* (chemically attaching them to a drug).

However, approximately 90% of all nuclear medicine

procedures are done using Tc-99m tagged / attached to an appropriate drug / pharmaceutical to be used for different procedures. That's one of the reasons that I call the radioisotope Tc-99m the Rolls Royce of all radioisotopes being used in nuclear medicine! It has the perfect energy (eV) and perfect physical half-life (T1/2).

There are two terms that must be defined:

Tagging and *radiopharmaceuticals.*

Tagging

Tagging is a series of processes done in a laboratory to chemically attach a *radioisotope* to any kind of *drug / pharmaceutical.*

It is done for the purposes of producing a specific *radiopharmaceutical* that is going to be used in a specific procedure.

What is a Radiopharmaceutical?

Radiopharmaceuticals (RP) are a combination of a *radioisotopes* tagged / attached to a *drug / pharmaceutical.*

The use of RP in nuclear medicine and PET scanning is like the use of *radio frequencies* in ultrasound imaging, *magnetic field* in MRI procedures, and *X-ray radiation* in radiography and CT scanning.

How Does a Radiopharmaceutical Find Its Way into Healthy or Unhealthy Tissues / Organs?

I would like to start this portion with a simple analogy, which I hope will help make this matter easier to understand.

Before cell phones, emails, texting, and, even way before

that, the creation of the post office, people communicated between long distances by way of *carrier pigeons* sending messages to specific addresses anywhere in the territory.

A message was tied to a pigeon's leg, and it was sent off to where the pigeon was familiar to and called it home.

It is also a fact that every element or chemical (drug) is attracted to a specific healthy tissue of an organ in our bodies. For example, iodine is absorbed by the thyroid gland, and glucose is extracted by healthy brain tissues. However, since not every radioisotope that is used in nuclear medicine procedures is able to find its way to the organ that we are interested in, we must find a way to carry / haul them into that organ.

Now, imagine that if we choose a specific substance / drug that is *attracted* to a specific organ, and attach / tag a *radioisotope* (the message) to it, while the drug finds its way to that organ, it will *haul* the radioisotope with it to that organ as well.

Then, during imaging, what is being detected by the gamma camera is the radioisotope and not the drug.

The drug (the pigeon) is the carrier and the radioisotope is the message.

NUCLEAR MEDICINE AND PET PROCEDURES

In nuclear medicine or PET procedures, radiopharmaceuticals (RP) are mainly extracted from the bloodstream by the healthy tissues of an organ and not by the cells or tissues that have been *altered* or rendered *nonfunctional*.

The altered tissues come in different forms of tumors,

angiomas, or metastasized tissues from other organs. Furthermore, RPs are not picked up by any artifacts such as bullets or surgical instruments left inside the patient's body by mistake.

For the sake of an example, let's say that we have administered Tc-Neurolite (*Tc-99m* being the radioisotope and the *Neurolite* being the pharmaceutical) to a patient for a brain scan. The Neurolite will be extracted by the brain's healthy tissues regardless of Tc-99m being attached to it. But what makes the imaging possible is the presence of radioactive Tc-99m in the organ and not the Neurolite.

After taking a series of images, if we see a uniform uptake / distribution of radiopharmaceuticals throughout the whole brain, it shows that there are no abnormalities.

However, if within the brain tissue there are any altered tissues, such as a tumor, hemangioma, metastatic tissues, or any artifact such as a bullet in the brain, no radiopharmaceutical / drug will be absorbed in the area(s). The unhealthy spot(s) will show as a cold / clear area(s). Finally, the size, shape, and location of the abnormality will be determined by the radiologist looking at a three-dimensional image of the affected organ.

By the way, since every minute area of our brain or other sensitive organs is dedicated to a certain function, it is crucial for the surgeons to know the *exact* location of the abnormality for their surgical purposes, and that is where the multimodality diagnostic procedures step in, such as SPECT-MRI or PET-MRI.

RADIOPHARMACEUTICALS USED IN NUCLEAR MEDICINE

These are the radiopharmaceuticals used in nuclear medicine procedures: Tc-99m, Tl-201, Ga-67, In-111, Xe-133, I-123, and I-131.

Tc-99m

It has a perfect half-life of six hours and the energy of 140 KeV. The sufficient half-life gives the nuclear medicine doctors and technologists enough time to perform all the procedural and post-procedural images, while the right energy level provides them with a sharp and clear image of the organ for a good and accurate diagnosis. Tc-99m's right energy also saves the patient from being overexposed to unnecessarily higher radiation.

Tc-99m is the radioisotope of choice for performing almost 90% of all diagnostic procedures in nuclear medicine. It is easily tagged with a variety of pharmaceuticals.

Tl-201

It has the half-life of three days (Tp is rounded off) and 70 KeV. It is a *potassium analog* radioisotope and has been the radiopharmaceutical of choice for all myocardial perfusion imaging for a long time. However, after the introduction of radiopharmaceuticals made with Tc-99m, Tl-201 is being used mostly for myocardial *viability tests*.

In 2008, the radiopharmaceutical Tc-Sestamibi was introduced for imaging of myocardial perfusion with less radiation exposure to the patient and better image quality.

Up to late 1980s to early 1990s, Tl-201 was being used

for *brain tumor imaging* as well, but that, too, became kind of obsolete with the development of a super quality imaging provided by PET scanning using F-18 FDG.

Ga-67

Gallium is an effective tumor-localizing agent. It has a T1/2 of three days, with 93 KeV. After the injection of Ga-67 to a patient, he will come back to the nuclear medicine department at 24, 48, and possibly 72 hours post-injection for imaging of mostly the upper body to determine the possible site(s) of the tumor(s) anywhere in the body.

In-111

Indium-111 is used in an indium scan to locate an unknown source of an infection in an unknown site anywhere in the body. It is a very time-sensitive and time-consuming procedure. Indium has a T1/2 of approximately three days, and the energy of 245 KeV. The procedure is called the *In-111 WBC (white blood cell) scan* or *Indium Leukocyte imaging*.

In this procedure, a timely follow-up of the protocol is very crucial; otherwise, the white blood cells will not survive and the *tagging* of the Indium radioisotope with a patient's white blood cells will not take place.

To start the procedure, 20 cc of the patient's whole blood is withdrawn in a heparinized syringe, and after a very detailed labeling of the syringe with the patent's information, it is sent to a radio-pharmacist. He will extract / separate the WBCs from the whole blood and will chemically *tag / attach* them to the radioisotope In-111.

Upon arrival, the *In-111 WBC* will immediately be injected back into the *same* patient and scanned at 4, 24, and possibly 48 hours post-injection. The white blood cells in radiopharmaceutical (In-111 WBC) are attracted to the infected area to fight off the infection. Therefore, where the In-WBC is concentrated, it will light up, showing the infected area(s).

Imaging is done as a whole-body scan because the source of the infection is unknown and could sometimes be found in the big toe.

Xe-133

Xenon is a radioactive gas with a T1/2 of five days and 81 KeV. It is used for pulmonary (lung) ventilation scans. After positioning the patient under or in front of the gamma camera, the patient is instructed to do a one-time bolus inhalation of the Xe-133 gas, and at the same time, a series of images of the lungs are being taken. If there is an area of the lung that Xe-133 cannot go through, there must be some kind of obstruction within the airways in that area.

Right after that, a lung perfusion scan with Tc-MAA (Macroaggregated Albumin) will provide the doctor with enough information to diagnose if the condition is an embolism (blood clot) in the pulmonary arteries or some chronic condition. With a lung scan, a chest X-ray is also needed to help the doctor with extra information. If the diagnosis is a pulmonary embolism (PE), an *immediate* treatment is required. PE can be a life-threatening condition.

I-123

I-123 comes in the form of a capsule, and the patient will take it orally. It is a diagnostic radiopharmaceutical for *thyroid imaging and function*. It has a thirteen-hour T1/2 and 159 KeV. For this test, the patient's *dietary preparation* is crucial, because the thyroid gland is a very small organ and any small amount of iodine in their diet, medication, or other previous medical procedures will have a drastic effect on the result of their test.

Not following the pretest guidelines will practically render the whole test useless.

Therefore, the dietary preparation for this test is very important. The patient will take the I-123 capsule orally, in the morning in the Department of Nuclear Medicine, and then go home. He is instructed to come back to the hospital at 6 and 24 hours post-capsule ingestion for scan and uptake.

When an I-123 scan is performed, *first*, several images will be taken to evaluate the *size*, *shape*, and presence of any possible *nodules* within the gland. *Second*, during the 6- and 24-hour imaging, a numerical value which is called *the percent uptake* is also calculated. This shows the amount of uptake of I-123 by the thyroid tissues.

The percentage uptake at the six-hour must be within 7–15%, and for 24-hour uptake must be around 10–30%. Any values lower or higher than these numbers will be an indication of *hypothyroidism* or *hyperthyroidism* respectively.

Any percentages within the normal values are considered *euthyroid,* or normal. Furthermore, if it is needed for the hyperthyroidism, doctors will use the size of the thyroid and the number of percent uptakes to calculate the *therapy dose of I-131* for that patient.

Also a 2 mci of I-123 capsule can be given to a patient for a whole-body scan to diagnose if the thyroid cancer has metastasized.

CHAPTER FIFTEEN

"Tough times never last, but tough people do."
—Robert H. Schuller

I-131 THYROID IMAGING AND THERAPY

I-131 is an interesting radioisotope with dual functions of *diagnostic* and *therapeutic* qualities. 10% of its decay is by gamma (ɣ) radiation with the energy of 365 KeV, which is used for diagnostic imaging, and 90% of its decay is by beta (β) particles with the energy of 606 KeV, used for therapeutic purposes.

In order to follow the thyroid procedures, a brief understanding of thyroid functionality will help.

Thyroid function depends upon a system of checks and balances between the *thyroid gland*, the *anterior pituitary gland,* and a section of the brain called the *hypothalamus*.

They turn each other on or off through hormonal level exchanges within the bloodstream.

The hypothalamus of the brain senses the level of T3 and

T4 hormones produced by the thyroid in the bloodstream, and depending on the level of these two hormones, the hypothalamus will release a hormone called Thyroid Releasing Hormones (TRH) into the bloodstream.

Then TRH will dictate the anterior pituitary gland to release Thyroid Stimulating Hormones (TSH), and this, in turn, will order the thyroid gland to produce more of T3 andT4.

A normal, healthy thyroid gland will extract *iodine* and *tyrosine* (an amino acid) from the bloodstream to make T3 and T4 hormones, which are needed in the bloodstream to *control* the metabolism of *oxygen* and *calories* into *energy*.

This is the way that any system of normal, healthy thyroid gland, anterior pituitary gland, and hypothalamus will work. However, if there are any miscommunications between these three elements because of the presence of tumors or nodules, the system of checks and balances will be off and, as a result, there would be either too little or too much production of T3 and T4 by the thyroid gland.

HYPOTHYROIDISM AND HYPERTHYROIDISM

Too little T3 and T4 hormones in the bloodstream will cause *hypothyroidism*, and too much of T3 and T4 will cause *hyperthyroidism*.

Symptoms of *hypothyroidism* include weight gain, dry skin, enlarged thyroid, fatigue, lethargy, hair loss, constipation, and slow heart rate. To treat hypothyroidism, the doctor will prescribe medication and the problem will be solved.

However, for *hyperthyroidism*, it is a different story.

Hyperthyroidism will cause conditions where all the symptoms are opposite symptoms of hypothyroidism, such as fast heartbeat, irritability, weight loss, bulging eyes, restlessness, diarrhea, mood swing, panic attack, and excessive hunger.

To treat hyperthyroidism requires more medical attention, but it *is not* a life-threatening condition at all.

Treating hyperthyroidism, which is caused by the autonomous and out-of-control thyroid cells making too much T3 and T4 hormones, can be done either by surgery (thyroidectomy) or a noninvasive treatment by I-131 radioisotope.

In some cases, both procedures of surgery and I-131 treatment are done in order to, first, get rid of the thyroid gland and second, wipe out the residual thyroid cells if there are any left behind.

Off the bat, I must say that I like I-131 radioisotope and it is one of my favorite radiopharmaceuticals, because it is a lifesaver. It also has a dual function of diagnostic and therapeutic.

There are 37 radioisotopes for the element *iodine*. The only stable one is I-127 and the rest are radioactive. However, in nuclear medicine, we only deal with two of them. The I-123, that is used solely for diagnostic purposes, and the I-131, which is used for both therapeutic and imaging.

Before we start talking about I-131 treatment, I would like to share a piece of information with all those people who someday might find themselves in the position of a patient and a candidate to receive I-131 treatment. The whole process of treatment is a piece of cake! Although it sounds and looks horrible and scary, it is the best and easiest way to deal with your condition. For those people who must go through I-131 treatment and might be a little nervous, let's go over a very simple walkthrough of all the diagnostic and therapeutic procedures in order to break down the barriers of the unknown false monster. A monster it is not, but in reality, it is an angel of cure and comfort!

After going through a simple diagnostic procedure, from the moment that your doctor decides to treat your hyperthyroidism with I-131, a barrage of restrictions, orders, and application signings will pour upon you and they all sound *super* horrible and frightening, but the good news is that all these restrictions are harmless and beneficial for your own protection. Some of these directions include: not eating anything that might contain iodine (iodine starvation is the goal) for a couple of weeks before your treatment; not getting pregnant before the test; not giving your milk to your infant child after the therapy dose you are taking.

When all these criteria are met, if the amount of I-131 for your treatment is less than 50 mci, you will ingest the therapy dose and will be sent home with a list of precautions. But if the amount of I-131 is over 50 mci, you will be hospitalized for two to three days before you can be released from the hospital.

If you are admitted to the hospital, your room itself looks very scary; everything in the room is covered, even the bedding, doorknobs, chairs, telephone set, toilet seat cover, floor, and some parts of the walls. Anything that you might touch, breathe on, or spit on is covered. It looks like you are being sent to the moon. During your hospitalization, family members can visit you from six feet away. Pregnant visitors and children are not allowed. Nothing will leave your room unless a trained technologist surveys it with his survey meter.

You will be instructed to flush the toilet twice every time you use the bathroom. There are also some dos and don'ts that you must follow.

Unfortunately, some of the nurses might be nervous around you because they don't know if they are pregnant themselves or they are trying to get pregnant, and that adds fuel to the fire of confusion and uncertainty.

Throughout your stay at the hospital, you will be cared for and monitored by doctors, technologists, and nurses. You also be able to have adult visitors as you wish.

Finally, after two or three days, you will be monitored and sent home with a long list of instructions.

The instructions will include directives about getting a ride home, living situation at home, workplace arrangement, trash management at home, and many more that all will benefit you and the people around you. Don't forget to sleep alone for ten to fourteen days. And, if you are a woman of childbearing age, make sure you are not getting pregnant for the next six months of post-therapy!

It all sounds horrible for someone who is going through a difficult time—but! Let me relate a personal experience. Don't worry! It all sound and looks horrible, but, as a matter of fact, that is all there is. It sounds and looks horrible, but in fact, all the paperwork you have been asked to sign is because they want to make sure that you understand the whole process and everything is consensual. Isolating you from the public during those times is because they want to protect other people from being exposed to any unnecessary radiation. And the reason for them to cover *everything* in your hospital room is to prevent any contamination coming from your urine, saliva, blood, and sweat, to those covered areas.

The reason behind it is because, after you leave the hospital, that poor nuclear medicine technologist must get down on his hands and knees to decontaminate the whole room (if you were naughty and did not follow the instructions during your stay). Everything must be decontaminated to the point that it is acceptable by the laws that are being enforced by the mighty NRC. Then the room will be ready for the next patient.

So, up to this point, all the actions were to protect you and the others. The more they inform you of the procedure, the better for you, your family, friends, and the public in general.

Listen to your doctor and the nuclear medicine technologist. They will provide you with a lot of information that's beneficial to you. They will ask you to drink a lot of liquids, which will cause frequent urinating that helps you to get rid of some of the background radiation from your body. During your hospital stay, they will provide you with some sour candies to suck on in order to salivate and flush out most of the unnecessary I-131 uptake by your salivary glands.

Remember that any type of liquid coming out of your body during these times is radioactive.

During your hospital stay, and at least three weeks post-ingestion of the I-131 capsule, you must follow three simple rules for protecting your loved ones and the public.

The three rules are: *distance*, *time*, and *shielding*. The farther you are from people the better, with a minimum distance of six feet apart. The less time you spend with them the better. And the thicker and denser the shield between you and the people the better (i.e. a concrete wall is better than a wooden door).

If you have noticed, it is all about protection and safety. The I-131 that you have taken will be concentrated in your thyroid tissues doing the job of slowing down or eradicating the hyperactive thyroid tissues at once. I-131 will do what a surgeon will do without the possibilities of complications, discomfort, and scarring.

It is a win-win situation for you and your family. It is also heartwarming to know that the success rate for I-131 therapy is 98%. For these patients, I-131 is an isotope sent from heaven!

The only side-effect is that the patient will develop *hypothyroidism* and require a replacement thyroid hormone by mouth indefinitely.

But remember, if you are the carrier of any kind of radiopharmaceutical in your body, you are ethically bound to follow the three rules of the *distance*, *time,* and the *shielding* until you are told that your body is free of radiation.

To be honest with you, during your stay at the hospital, there is something that you might be complaining about, and that is *boredom*. Don't forget to take a couple of good books to read and your cellphone and the charger, because there is only so much TV one can watch!

Good luck to those who would be a candidate for getting the I-131 therapy for their hyperthyroidism or thyroid cancer. And remember! The whole process might look and sound horrible, but it is so easy, and you are getting rid of something that your body does not need!

CHAPTER SIXTEEN

"Pollution is nothing but resources we're not harvesting."
—Buckminster Fuller

FREQUENTLY DONE NUC-MED PROCEDURES

There are many nuclear medicine procedures being done every day; however, there are a few that are done more frequently than others.

My hope is that this section of the book will shed some light on the unfamiliar nuclear medicine procedures for those patients who are a bit nervous about going through them. These procedures include cardiac tests, bone scans, lung / VQ scans, liver scans, brain scans, and renal scans, to name some of the most intimidating ones.

Furthermore, I must admit that the name "nuclear medicine" itself is not a patient-friendly name at all!

It is easy for doctors and technologists to be comfortable with the name. However, for a patient who is uneasy with the whole situation, adding the name "nuclear" must be frightening!

For that reason, I would like to propose a new, patient-friendly name to replace the old scary name. I would like to call it: "The Department of Radiopharmaceutical Imaging (RPI)."

Well, who is going to listen to me?

Let's get on with the business of explaining some nuclear medicine procedures.

RADIOPHARMACEUTICAL PREPARATION

As we learned earlier, nuclear medicine procedures are possible only by using *radiopharmaceuticals,* and every radiopharmaceutical must be prepared for each and every specific procedure.

Usually, radiopharmaceuticals are prepared in three ways:

1. The imaging is done by targeting the radiopharmaceutical into the *healthy tissues* of an organ.

2. Targeting the radiopharmaceutical into the *altered tissues,* such as *tumors, hematoma,* or *metastasized tissues,* that are not functioning as a healthy tissue of that organ.

3. Radiopharmaceuticals that are prepared to determine the *patency* of the organ's plumbing, such as the biliary tract in the liver, the airways, and the pulmonary arteries of the lungs, or the coronary arteries nourishing the myocardium.

In the patency procedures of different organs, if the radiopharmaceutical is stopped at any point within the plumbing (the passage of blood, bile, or air), that's the problem point where something must be done.

For the first and second techniques of preparing the radiopharmaceuticals, I would like to explain the *targeting technique* by a simple analogy.

Imagine you have a piece of fabric, and you can cut it into the shape of a liver, heart, brain, or any other organ in the body. Then, make a hole the size of a one-dollar coin somewhere on the fabric. Get a piece of *magnet* the size of the one-dollar coin, place it at the hole you made in the fabric, and secure it with a piece of tape. To complete the experiment, you need a bowl of water and a can of iron filings.

Let's assume that in this experiment, the fabric is the healthy tissues of the organ, and the coin-sized *magnet* in the middle of it is the altered tissues or the tumor.

First, we throw the bowl of water on the fabric and the magnet. What will happen? The fabric will absorb the water, while the magnet will not absorb any water at all.

Continuing with a second experiment, if we throw the whole can of *iron filings* onto the fabric and the magnet, and shake the fabric off, all the iron filings will fall off the fabric, but the magnet will hold on to those iron filings.

In the first experiment, the radiopharmaceutical was targeted to the healthy tissues, and in the second experiment, the radiopharmaceutical was targeting the tumor or non-functioning cells within the same organ.

As a rule, where the radiopharmaceutical is absorbed,

it will show up in the image as an uptake, and where the radiopharmaceutical was not absorbed, there will be no activities and is technically called a "cold spot."

You don't even have to be a doctor to look at these images and say, "Oh, look at that!" The results will be obvious to everyone.

MYOCARDIAL PERFUSION SCAN

Myocardial perfusion scan is a combination of a *patency* test of the *coronary arteries* and targeting of the *healthy myocardium*. If the coronary arteries (the arteries that supply blood to the heart muscles) are blocked off by *plaque* (cholesterol and fatty deposits), the myocardium being supplied by that coronary artery would not be receiving any blood, and in turn, that part of the heart will not show up in the image. Eventually when the blockage is severe enough, the myocardium beyond the point of blockage will die or infarct.

The other *patency* test is the *hida scan,* when the *biliary tract* within the liver, at some point, might be blocked because of *inflammation* or a *stone.*

The *lung ventilation scan* also follows the same principle if the airways are blocked off for any physical reason.

Also, in a *lung perfusion* scan, when the flow of blood is stopped by an *embolus* to any of the lung's lobes, that lobe is not getting the radiopharmaceutical and it does not show up in the scan.

CHAPTER SEVENTEEN

"The butterfly counts not months but moments, and has time enough."

—Rabindranath Tagore

CARDIAC TESTS

Noninvasive diagnostic cardiac tests are done for a variety of reasons; however, a big chunk of it is because of the coronary artery disease (CAD).

Our heart muscles (myocardium), like every other muscle in our bodies, need a good supply of oxygenated blood flow for oxygen and nutrients. There are two main branches of arteries (coronary arteries) supplying oxygenated blood to the myocardium: the *Right Coronary Artery* (RCA), and the *Left Coronary Artery* (LCA). The heart muscles, like any other muscles in our bodies, when they are overworked and deprived of oxygen and nutrients, will cramp up, much like the leg cramp of a tired soccer player or a marathon runner.

When, for any reason, either or both coronary arteries

are narrowed or blocked off, the condition is called *atherosclerosis*, and the blood supply to those areas of the myocardium will be diminished or stopped, which will cause the myocardium to cramp up, or as the medical professionals call it, *angina*.

ANGINAS

Angina is not just a mild, simple chest pain. It is a major pain, pressure, tightening, burning, elephant-like weight on your chest. Sometimes, along with angina comes pain in the left arm, neck, jaw, shoulder, or back. Furthermore, along with it, one might experience dizziness, fatigue, nausea, shortness of breath, or sweating.

There are two types of anginas:

First, *stable angina,* which happens with exercise, goes away with resting and / or taking your heart medication, and lasts less than five minutes. However, beside the exercise, there are other conditions that might trigger angina, such as emotional stress, smoking, heavy meals, or cold temperatures. When this happens, it is a good idea to call your cardiologist soon.

Second, *unstable angina*, which happens even at rest, lasts longer than thirty minutes, and rest and medication do not help. In this case, a heart attack may occur.

Unstable angina is a 911 situation!

Before you experience any type of *angina*, it is advised to stop smoking, control your blood pressure, lose the extra weight, exercise regularly, watch your cholesterol and diabetes, and pay attention to your family health history (the family health history is one of the most important risk factors).

WHAT CAUSES ANGINA?

Besides having a history of family cardiac problems, having an at-risk lifestyle in terms of bad diet, lack of exercise, and some other health conditions can cause the buildup of fat and cholesterol at the inner wall of the coronary arteries, which will form plaques.

These plaques are not soft but very rigid, and over time, they will cause the narrowing of the inner diameter of the coronary arteries, causing slower blood flow. If the condition is not taken care of, these plaques will grow bigger and eventually completely block off the passage of the blood to the myocardium beyond that point, causing it to die (infarct).

At this point, two scenarios might take place.

First, the myocardium is damaged beyond repair and is dead (infarct).

Second, the myocardium is still *viable,* and it can be saved.

CARDIAC PROCEDURES POST-ANGINA

⇒ First, a Tc-Sestamibi scan will be performed in order to evaluate an overall condition of the heart muscles (myocardium), to see if there is any section of the heart that is not getting enough blood.

⇒ Next, if the Sestamibi scan is positive, then a thallium viability test (Tl-201) will be ordered by the physician to find out the extent of the damage to the myocardium and its viability. If it is confirmed that the myocardium is viable, then the patient might go through a few more procedures as follows: Coronary angiography to find the exact location of the atherosclerosis (blockage in this case), by using a radio-opaque contrast under the X-ray fluoroscopy.

⇒ Angioplasty. When the patient is under the X-ray fluoroscopy and the exact location of the blockage is confirmed, the angioplasty will be performed by inflating a balloon at the narrow passage to widen the inner diameter of the coronary artery to resume the normal flow of blood to the myocardium below that point.

⇒ Atherectomy. Again while the patient is under the X-ray fluoroscopy, if the procedure of angioplasty is not enough to widen the passage, then an atherectomy procedure is the next option. It is a procedure involving a rotating blade to shave off the plaques inside the coronary artery. The shaving of the plaques makes the inner diameter of the artery wider for a proper passage of the blood to the myocardium below that point.

⇒ Coronary artery stent placement. While the patient is still under the X-ray fluoroscopy, a stent—which is a wire-meshed tube—is placed at the narrowed area to keep it open for a normal flow of the blood. It is a procedure done before CABG (described below).

⇒ Coronary Artery Bypass Graft (CABG). I don't believe the cardiac surgeons are doing as much of

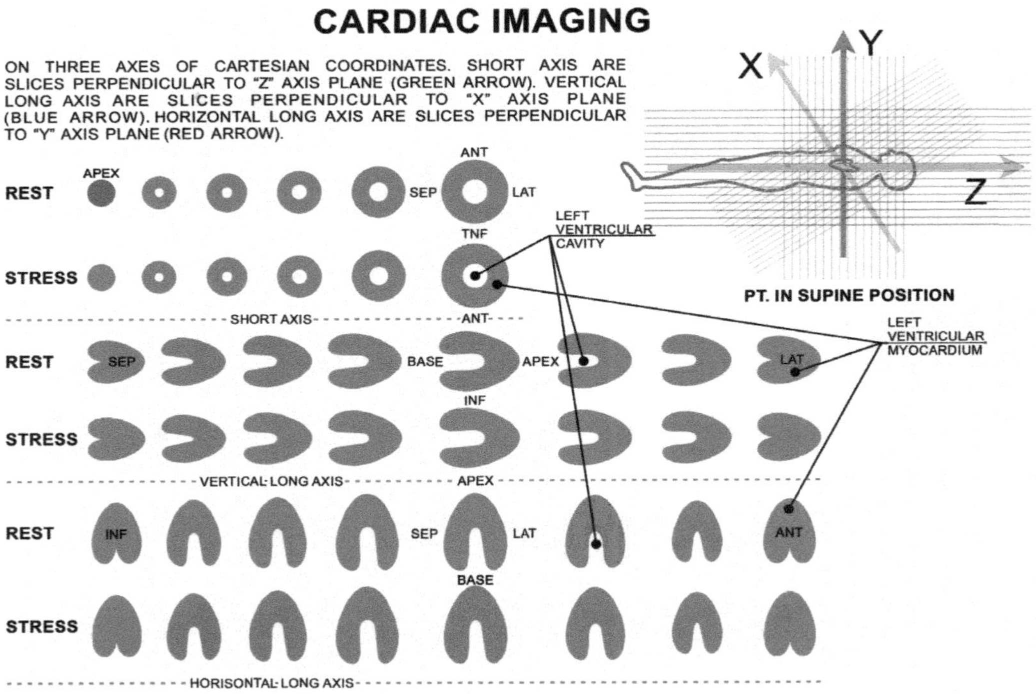

CARDIAC IMAGING

ON THREE AXES OF CARTESIAN COORDINATES. SHORT AXIS ARE SLICES PERPENDICULAR TO "Z" AXIS PLANE (GREEN ARROW). VERTICAL LONG AXIS ARE SLICES PERPENDICULAR TO "X" AXIS PLANE (BLUE ARROW). HORIZONTAL LONG AXIS ARE SLICES PERPENDICULAR TO "Y" AXIS PLANE (RED ARROW).

PT. IN SUPINE POSITION

LEFT VENTRICULAR MYOCARDIUM

LEFT VENTRICULAR CAVITY

REST
APEX
SEP
ANT
LAT
INF

STRESS

- - - - - SHORT AXIS - - - - -

REST
BASE
ANT
APEX
SEP
INF

STRESS

- - - - - VERTICAL LONG AXIS - - - - -

REST
APEX
SEP
LAT
BASE
INF

STRESS

- - - - - HORISONTAL LONG AXIS - - - - -

Fig. 27

this type of invasive cardiac surgeries as they have in the past. In bypass surgery, surgeons will cut out a piece of vein from the patient's leg or a piece of artery from his arm, and surgically attach it to both sides of the blockage point. It will bypass the blockage area to restore the normal flow of blood to the heart muscle below that point.

MYOCARDIAL PERFUSION SCAN (REST AND STRESS SCAN)

The reason for your doctor to request this test for you can be any or a combination of the following reasons:

• If you have complained of a *chest pain* and your doctor thinks you might have had a *heart attack.*

• Assessing your heart function after a *bypass surgery*, *angioplasty*, or *stent placement.*

• Or just looking for the advancement of your cardiac artery disease (CAD).

Myocardial perfusion scan is composed of two parts.

RESTING SCAN

Resting scan is done after the administration of a radiopharmaceutical while the patient is at rest. Then a set of all-around imaging (cine) of the heart is done, through a process called Single Photon Emission Computerized Tomography (SPECT). Right after the completion of the *resting scan,* the patient is ready to go through a *stress test.*

STRESS SCAN

There are two ways to do a stress test.

First, if the patient is able to run on a treadmill or paddle on a stationary bicycle for about 6–8 minutes, and his heart rate reaches at least 85% of his predicted maximum heartrate (patient's age minus 220 is a number that is equal to 85% of his predicted maximum heartrate), the attending physician will ask the technologist to administer the radiopharmaceutical, and that would be the end of the stress test. Later, the patient will go for the second set of imaging for a comparison of resting and stress images.

Second, if the patient is unable to perform any kind of physical activity, the stress test will be done *pharmacologically*. It is called an *adenosine stress test*. Adenosine is a vasodilator drug causing the dilation of the vascular system; it causes a drop in the patient's blood pressure, and as a result the patient's heartrate will increase, mimicking exercise. The administration of adenosine is not a bolus injection, but through an automatic pump over the course of a few minutes. Then, when the patient's heartrate reaches 85% of his maximum heartrate, the attending physician will ask the technologist to administer the radiopharmaceutical (Tc-Sestamibi) and that would be the end of the stress test. A few hours later, he will be ready for imaging.

Sometimes during the adenosine stress test, the patient's heartrate does not reach 85% of his maximum heartrate, and he will be instructed to move his feet and fingers minimally to bring the heartrate up, and usually that will do the trick.

Then both sets of the rest and stress images will be displayed in three different cuts of *short axis*, *vertical long axis,* and *horizontal long axis* for a comparison of the myocardium from all around the heart during two conditions of rest and stress. These images will reveal if there are any area(s) of the heart that are not receiving enough blood and nutrients in rest and more so during the exercise. If there are any defects shown in the images, then the doctors will see which coronary artery is being blocked off or otherwise unable to deliver blood to those questionable areas.

MUGA (MULTIGATED) SCAN:

When a doctor is suspicious of CHF (congestive heart failure) concerning the pumping ability of their patient's heart, he will prescribe a MUGA test.

To start a MUGA test, about 3 cc / ml of the patient's whole blood is withdrawn in a 5-cc heparinized syringe, then injected into a *cold kit RBC* (red blood cell) vial. The blood is swirled gently, then receives about 30 mci Tc-99m, after which it will rest for twenty minutes.

Then your radiopharmaceutical of Tc-RBC is ready to be injected intravenously back into the *same* patient.

After injecting the Tc-RBC, we have three hours of optimal time to scan the patient.

To start the scan, the patient is placed on the table. Three EKG electrodes will be peeled off and stuck on patient's right shoulder, left shoulder, and below the left rib cage.

All three wires are attached to the gamma camera, and hopefully we will see a beautiful EKG showing on the monitor. In some rare cases, the rhythm of the patient's

EKG and the gamma camera's EKG are not synchronized, and therefore, the imaging will not be possible. Sometimes this happens because the patient has an irregular heartbeat or is too nervous. Now comes the technologist's expertise to calm the patient down and make him more comfortable while moving the three electrodes around on the patient's body. In male patients, shaving the chest hair where the electrodes are attached will make a big difference. If none of these tricks work, you may have to make some changes in your gamma camera's settings.

The purpose of connecting the patient's EKG to the gamma camera is to obtain a dynamic video of the heartbeat in three positions of anterior, left anterior oblique, and left lateral view.

After the whole procedure is done, there would be three moving images of the left ventricle's *cavity* being *contracted* and *expanded* (depolarizing and repolarizing), showing how the patient's heart is pumping blood out (ejection rate).

By looking at these moving images from three different angles, doctors can tell the *size* of the *left ventricular cavity* and the left ventricle's *wall motions*.

If there are any parts of the *ventricle's wall* that are damaged because of ischemic or infarct tissues, in those sections, there would be no motion or very little movement.

By the way, *Ejection Fraction* (EF) is also another piece of information provided by the computer after a MUGA scan. EF is a volumetric fraction of blood ejected out into the aorta from the left ventricle with each heartbeat. *EF is a comparison of the amount of blood leaving the ventricle to the amount of blood that remains in the ventricle.*

Ejection fraction is calculated in a percent value.

If it is between 55–65%, the EF is normal.

40–55% the EF is mild.

30–40%, the EF is moderate.

Less than 20%, the EF is a sign of severe CHF.

However, many physicians believe these numbers are not a true representation of the patient's condition and the referring physician must pay attention to how the patient feels and is doing, rather than these raw percentages in front of them.

Note: In MUGA scan, the images that the doctors are looking at are the left ventricle's cavity expanding and contracting and not the myocardium itself.

CHAPTER EIGHTEEN

"The universe is not outside of you.
Look inside yourself,
everything that you want,
you already are."
—Rumi

LIVER SCAN AND HIDA SCANS

In these two different scans, the liver is the target organ. However, the radiopharmaceuticals being used for these two procedures follow different paths. For the liver scan, the radiopharmaceutical is used to distinguish between the *healthy* liver tissues vs. *unhealthy* liver tissues, while in a HIDA scan, the radiopharmaceutical is used to show the *patency* of the *hepatobiliary tract* (the plumbing of the bile within the liver).

In *liver scanning,* the radiopharmaceutical Tc-SC (technetium sulfur colloid) is used, which will be extracted from the blood-pool by the healthy functioning liver tissues. The unhealthy or altered liver tissues, such as tumor, hematoma, or metastasized cancer cells from other organs, would not be able to extract the Tc-SC, and as a result those areas will show up as cold spots (lighter / clear). Not all abnormal spots in liver scans are cancerous; however, further testing such as *biopsy* is needed to determine the makeup of the questionable spots / tissues.

HIDA SCAN

A HIDA scan is done if there is any question of obstruction of *bile* movement anywhere throughout the *biliary tract,* or the *gallbladder* itself. The obstruction can occur because of the presence of stone or inflammation. The radiopharmaceutical *Tc-mebrofenin* is administered intravenously, and the imaging will start immediately after the injection. A series of images are taken with the intervals of 5, 10, or 15 minutes for up to one hour to show the movement of the radiopharmaceutical throughout the biliary tract.

The biliary tract within the liver is just like a stream in the mountains, collecting rainwater into a river and taking it to the sea. The biliary tract will collect the *bile* throughout the liver, storing some of it in the *gallbladder* for a later use, and some of it will be released into the *duodenum* through the *common bile duct* for digestion of food.

A little more detailed look at the bile collection within the liver shows that at the beginning of the biliary tract, there are two main branches of *right* and *left hepatic ducts.* Further down the road, these two will join together to form the *common hepatic duct.* Further down, it will bifurcate into the *cystic duct* (going to gallbladder) and the *common bile duct* (going to duodenum).

If all goes well and there is no obstruction, after an hour or so, all of the radioactive tracer will be emptied into the duodenum, and the liver image will disappear from the monitor, and all we will see is the radioactivity in the small intestine.

Now, what if there is an obstruction somewhere along the way? There are three main areas where the obstruction can occur.

If the obstruction is within the *common hepatic duct*, then the liver will not disappear, and the gallbladder, the common bile duct, and the duodenum would not show up. On the other hand, if everything else shows up and the gallbladder does not, the obstruction is in the *cystic duct*. If the gallbladder and the common bile duct show up and there are no activities in the duodenum (bowel), then the obstruction must be somewhere within the *common bile duct*.

It is very simple reasoning to determine the site of the obstruction. Sometimes the obstruction within any section of the biliary tract is not as severe and the use of some interventional drug will show us the severity of the obstruction. One of those drugs is CCK (cholecystokinin), which is used when the gallbladder is full of radiopharmaceutical, but it does not empty into the common bile duct and keeps getting bigger. If the obstruction is not that severe, the intravenous administration of CCK will stimulate the release of bile (radiopharmaceutical) out of the gallbladder, and if the obstruction is severe, there would be no changes. Furthermore, if the gallbladder isn't showing up and all the radiopharmaceutical is being emptied into the *duodenum*, then the intravenous administration of *morphine* will stop the flow of the radiopharmaceutical into the duodenum, and instead pushes the radiopharmaceutical into the gallbladder. If the gallbladder shows up, the obstruction is not that severe, and if it does not, there is an indication of a major obstruction in the cystic duct or the gallbladder itself.

So, this procedure is one of those procedures that shows the patency of the biliary tract's plumbing.

V/Q SCAN (LUNG SCAN)

There are a few emergency procedures that are done in nuclear medicine, and V/Q scan is one of them.

V/Q scan is a lung scan requested by a physician when a patient is experiencing shortness of breath, and there could be a possibility of pulmonary embolism (PE).

Pulmonary embolism is a life-threatening condition, which must be dealt with immediately!

Shortness of breath can be caused by a blood clot (embolus)

in the lung. An embolism can cause a partial or total blockage of blood flow (vascular occlusion) in a section(s) / lobe(s) of the lung.

In this situation, shortness of breath is not the only worry; it also can cause heart attack or stroke if the embolus breaks away.

That's why, even if a pregnant woman is in this situation, the physician does not hesitate to ask for a V/Q scan.

If a V/Q scan is ordered on a pregnant patient, the regular doses are cut down in half in order to protect the fetus. That's why I believe Tc-99m is the Rolls Royce of all radioisotopes for its perfect, efficient low energy and its short half-life. It can even protect a fetus!

Anyway, how the V/Q scan is done is very easy.

The ventilation part is done first, and that is by placing a special airtight mask on the patient's mouth and nose, and after the radioactive gas (Xe-133) is released inside the mask, the patient is instructed to take *one* deep and heavy inhalation of the gas and hold it in for approximately five seconds, then breathe normally for a minute into the same mask. This concludes the ventilation part. During the ventilation, from the beginning, the gamma camera is on, taking a dynamic video of the whole procedure. The second part will be the lung perfusion scan, which will be done immediately after the ventilation part.

For a perfusion scan, the patient must be lying down on his back for a uniform distribution of Tc-MAA (Macroaggregated Albumin) in the lungs. Imaging will be started at five minutes post-injection. Images will be taken from different angles and positions such as anterior, posterior, right and left laterals, and all the four obliques around the right and left lungs.

V/Q scan is a procedure where, at first, the patency of the airways of the lungs is examined, and second, the patency of the lung's vascular system is being tested.

With this test, a chest X-ray is also necessary!

BONE SCAN

Bone scan is done more frequently in nuclear medicine than any other scan.

It is done for a variety of reasons, including:

- Bone infection (osteo-myelitis)
- Degenerative bone disease (osteo-arthritis)
- Bone tumors, which could be *benign* or *malignant* (osteo-sarcoma)
- Bone stress fractures
- Or if there is a question of metastasized cancerous cells from other organs, such as lung cancer

Bone scan is done with the intravenous injection of the radiopharmaceutical Tc-MDP to obtain a whole-body imaging of the skeletal system, two to three hours post-injection.

When there is any bone cell alteration, the *osteoblasts* try to heal the condition by forming new bone cells. Where the osteoblasts are active (the new cells), there will be a higher concentration of radiopharmaceutical Tc-MDP and it will manifest itself as "hot spot."

The good news is that not all hot spots are cancerous, but it tells your doctor to look back into your past medical records, and possibly order some other exams such as CT, MRI, biopsy, or just a simple bloodwork to come up with a more concrete understanding of your condition.

BRAIN SCAN

There are a few modalities and multimodalities that perform brain imaging, and some provide the doctors with excellent detailed images of the brain. However, nuclear medicine and PET brain imaging are superior to others in terms of providing doctors with *functional* aspects of the brain. For example, when there is a question of *brain death, epilepsy*, or *Alzheimer's disease*, these two modalities are way ahead of the pack.

In brain imaging, the availability of two different kinds of radiopharmaceuticals gives the nuclear medicine imaging or PET imaging the choice to either target the healthy tissues or the unhealthy tissues, such as tumors.

RENAL SCAN

Kidneys are our body's main filtration system. Approximately 150 quarts (37.5 gallons) of blood pass through our kidneys every day, producing 1 to 2 quarts of urine daily. This is done to balance the fluid level in our bodies, to remove the waste products in the blood, to regulate the blood pressure, and to regulate the electrolytes in our blood (e.g. sodium, potassium, calcium, magnesium).

Therefore, whenever there are any physical or bacterial obstructions to the kidneys and its appendages, any of these problems may occur:

- Renal artery stenosis, or the narrowing of renal arteries.
- Hydronephrosis, which is the dilation of the *renal pelvis* due to urinary tract obstruction.
- Hydroureter, the dilation of the *ureter*.
- Hydronephroureter, the dilation of both *renal pelvis* and *ureter*.
- Pockets of infection or abscesses.
- Hypertension.

To diagnose these conditions, there are a few diagnostic modalities such as ultrasound and CT. However, the nuclear medicine *renal function scan* is one of the most helpful procedures because it shows the functionality of both kidneys at two levels of blood flow, and urine formation and excretion of urine through urinary tract.

The nuclear medicine *renal function scan* is a one-hour procedure where a complete set of images are taken.

The *first stage* right after the intravenous injection of the radiopharmaceutical (Tc-MAG-3), is a series of dynamic images taken to show the patency of all renal arteries and the flow of blood through the kidneys.

The *second stage* of renal function is the *formation* and the *excretion* of urine through the renal pelvises to ureters, and finally into the bladder.

During this procedure, any obstructions and abnormalities at any point will be recorded. Sometimes only one of the kidneys will show up, although anatomically both kidneys are there. This test even provides us with a percentage of each kidney's function. For example, 85% right kidney and 15% left kidney.

Furthermore, this procedure is requested for the evaluation of kidney transplants.

It is a noninvasive procedure with a very low exposure to radiation because right after the test, almost all the radiopharmaceuticals leave the body by urinating, which makes it one of the shortest *biological half-life* exams.

PET SCAN

I was lucky enough to meet Professor Michael E. Phelps, who is often credited with inventing the PET gamma camera.

I met him around the early 1990s at UCLA when one of the first PET scanners was being installed in the Department of Nuclear Medicine. I found him a humble and down-to-earth man who did not put himself on a pedestal for creating one of the most amazing inventions in the history of nuclear imaging.

The PET gamma camera is an invention that shed a bright light onto the diagnostic procedures of two of the most mysterious illnesses that human beings have been wrestling with for many years: epilepsy and Alzheimer's disease, alongside other sensitive procedures.

PET scanning can show the exact whereabouts and the severity of these diseases.

For this wonderful invention, Professor Phelps was awarded with some of science's most prestigious honors. He was given the Enrico Fermi award and was appointed to the National Academy of Science in 1998. He was also awarded the Massry Prize from the Keck School of Medicine of USC in 2007.

PET scanning is also helpful in prognosis and diagnosis of heart disease and metastasized cancerous tissues anywhere in the body.

Nuclear medicine and PET scanning both use gamma detector cameras. However, there are two major differences between the two.

First, the nuclear medicine cameras consist of two detectors, while the PET scanners have an all-around doughnut-shape detector.

Second, in nuclear medicine, the radiopharmaceuticals are gamma emitters, while for PET imaging, the radiopharmaceuticals are beta (positron) emitters.

What makes PET scanning possible is very interesting. In nuclear medicine procedures, the radiopharmaceuticals are gamma emitters and ready to interact with the gamma camera detectors. However, in PET procedures, right after the emission of positrons, each positron will collide with its *antiparticle,* which is an electron, and they both *annihilate* into a pair of gamma radiations traveling in two opposite directions. Then both are registered in two opposite sides of the doughnut-shaped detector.

By the way, each of these gamma radiations possess 511 KeV energy traveling with the speed of light.

F-18 FDG is a positron emitter radiopharmaceutical for investigating neurological problems such as epilepsy and Alzheimer's disease.

PET scanning reveals that the more active areas of the brain during an epileptic seizure will consume more glucose and will show up as a "hot spot," while in the case of Alzheimer's disease, where there are less brain activities, there would be less concentration of F-18 FDG and it will show up as a "colder spot."

Furthermore, as a rule, it is known that radiopharmaceuticals will concentrate more in tissues that are multiplying faster, and since cancerous tissues (tumors) are multiplying more rapidly than the normal functioning tissues, there would be more concentration of radiopharmaceuticals in those areas and will show up as a "hot spot."

Note: the element fluorine has only one stable atom, which is fluorine 19 (9 protons and 10 neutrons in its nucleus), and no natural *isotopes*. However, there are 18 radioisotopes of fluorine that are made synthetically in the laboratories. Seventeen of them have a very short physical half-life ($Tp = 10^{-9}$ second), but the isotope F-18 with the longest Tp of 110 min is the most suitable isotope being used in PET scanning. Fluorine has 9 protons in its nucleus, and in order to emit beta particles (positron), it will go through an electron capture (εc) decay, losing one proton (p) and gaining one neutron, transmuting into a stable atom of oxygen with 8 protons in its nucleus.

In recent years, the technology of *multimodalities* such as CT/PET and MRI/PET have been utilized for a better *localization* of the affected area. CT and MRI will show a clear *anatomical* view of the organs, while the SPECT and PET scanning show the *functionality* of the tissues. Thus, the overlapping of the two views of anatomical and functional views of an organ will pinpoint the exact location of the affected area, making it easier for further procedures.

CHAPTER NINETEEN

"Only those who dare to fail greatly can ever achieve greatly."
—Robert F. Kennedy

NUCLEAR REGULATORY COMMISSION

In 1946, after WWII, President Harry S. Truman created the Atomic Energy Commission (AEC), turning the control of *atomic energy* from the military to civilian hands. Later, in 1974, the *Reorganization Act* established the Nuclear Regulatory Commission, assuming the AEC's responsibilities. NRC is responsible for protecting public health and safety, environmental quality, national security, licensing, regulating the use of medical radioisotopes, and the nuclear energy facilities that generate electric power.

The NRC is headed by five commissioners appointed by the president and confirmed by the senate for five years.

The commission is headed by a chairman whose actions are governed by the general policies of the commission.

The commission is a collegial body that formulates policies and regulations, governs nuclear reactors and nuclear material safety, and handles all the licensing.

Now we know that NRC is the *boss* when it comes to radioactivity and the way it ought to be handled.

Nuclear medicine and the PET are two modalities that use radiopharmaceuticals of different radioisotopes in different amounts. The fact that radiopharmaceuticals are odorless, colorless, and their volumes have no indication of what type of radiopharmaceutical we are dealing with makes them impossible to tell the difference, unless they are *religiously labeled*.

Another important factor to be considered is that these radiopharmaceuticals are being administered to a human being, where any mistakes would be emotionally and economically very costly. Therefore, the NRC has a huge responsibility to regulate and oversee the implementation of all its regulations in every step of the way.

Radioisotopes, from their production in nuclear plants or cyclotrons, to nuclear laboratories, hospitals, and departments of nuclear medicine, are under the watchful eyes of the NRC.

Furthermore, administering a radiopharmaceutical to a patient and completing a procedure is not the end of their responsibilities, but it is the beginning of the patient's education. They should be informed how to behave at home and in public and the proper storage of the things they use that can be contaminated with radioactivity.

Now, how does the NRC know who is doing what? The only way they can control the situation is through a periodic, thorough inspection of our record-keeping at our facilities. And I mean they go through our record-keeping with a toothpick and comb! They do not leave any stone unturned!

During these inspections, everyone in a facility—including the department heads—better be there and ready. The NRC inspectors are not going to sneak up on you and surprise you. There are quarterly or annual inspections that they will announce way ahead of time. You will know the exact date and time of their inspection, and they will *show up* on the hour.

You must be there, ready to answer many questions! Remember, if they are not happy with what they hear and see, your license may be suspended or revoked!

After all, they are not there to crucify you. They are there to make sure everything is being done the way it should be done. They are there to teach you all they know in order to make things better. Seize the moment and learn as much as you can from them. It is better for you, your institution, your patients, and the mighty NRC itself!

OCCUPATIONAL RADIATION EXPOSURE

One of the main branches of the NRC's responsibilities is how radiopharmaceuticals are used in medical facilities. It is one of the most sensitive aspects of their responsibilities because of the involvement and the welfare of the patients and the medical personnel as well.

In a medical facility, there are two groups of people who are directly exposed to radiation and its byproducts.

First, the medical personnel.

Second, the patients and their families.

First, the hospital staff such as doctors, technologists, and nurses who are in direct contact with "hot" patients. They all must be wearing *radiation badges* to monitor their work practices, to see if they are following the radiation exposure guidelines set by the NRC. At the end of each month, badges are collected and sent to a lab to be analyzed. If any of these badges show overexposure, a notification will be issued to the facility management for the correction of the staff's work habits and practices.

The guidelines for the allowable amount of radiation exposure to each employee for each month is very specific and complex. The exposure level is different for every organ in the body. For example, our eyes are more radio-sensitive than our hands, and in general, any organs in our bodies that are more active and produce new cells such as bone marrow are more sensitive to the radiation than those cells that are less active.

Second, the patients and their families who must be educated enough to know how to behave around each other, friends, colleagues, and the public in general when they leave the medical facilities. And again, they must understand the three-prong rules wherever the patient goes: *distance*, the longer the better; *time*, the shorter the better; and *shielding*, the thicker the better.

And last, the proper way of disposing or storing radioactive waste is by the patient themselves at their homes.

One more thing! Please change the name of the nuclear medicine to *Radiopharmaceutical Imaging (RPI).* It sounds better and friendlier to our patients!

USE OF RADIOACTIVE MATERIALS AND RADIATION IN OTHER INDUSTRIES

I must mention that radioactive materials are not used exclusively in medical facilities, but also in just about all other types of industries. Radiations are used in airports and border control for security and inspections. In metal industries, it is an easy and cost-effective quality control for defects and consistency of the parts.

EPILOGUE

At last!

Upon finishing this book, I learned a new language: the language of the universe, which made me understand the interconnectivity and harmony that exist among us and the whole universe. We have a kinship with all the stars in the universe.

Now I understand that it has been a long journey for us to get here, from the Big bang to galaxies, stars, and finally to our one and only habitable celestial system, which is our home that we call the solar system. What a journey! A journey that took the Mother Nature 13.8 billion years to perfect. Now, the question is: Do we appreciate such a precious gift that has been handed to us by whomever or whatever that we believe in? I believe that for the last 300 years, since the beginning of the Industrial Revolution, we have been chiseling down the very foundation of our only sanctuary in the whole universe.

My hope for us, for humanity, is that we will last for

at least another five billion years before our sun falls into its disastrous red giant cycle, getting so big as to engulf Mercury, Venus, and our beautiful Earth, finally becoming a *planetary nebula,* wandering in space until we are adopted by another loving galaxy.

Nowadays, when I read the poems by Omar Khayyam, the Persian poet, philosopher, mathematician, and astronomer, I get a taste of the old world's philosophy and the new world's scientific findings. As Omar Khayyam philosophized, the cyclical nature of life and death, "dust into dust," mirrors the most contemporary astronomical beliefs of the galactic life cycle, "nebula to nebula."

INDEX OF FIGURES

A NOTE FROM THE AUTHOR

The United States of America is a furious inferno!

An inferno with no mercy, melting everything and everyone that is thrown into it on a fast-moving conveyer belt.

Everyone and everything is treated the same. Of course, exceptions do exist all over the world.

If you find yourself on that conveyer belt as I found myself, it doesn't matter if you were born in a poor neighborhood in another country with no knowledge of the English language, or if you were born right here in this beautiful land. You will be crushed, crumbled, and melted into molten ready to be poured into a mold.

There are thousands of different molds, molds that will shape you into a great leader or a failure. Thank goodness the choice is ours, going through a treacherous roller coaster of our daily lives.

I was born in Iran and came to this country at the age of twenty-four. I remember, the second day in this country, I started working in a gas station pumping gas, wiping windshields, and checking the motor oil, radiator water, and air pressure in tires. All these services and courtesies were given for buying a gallon of gas (at only 25 cents a gallon!). I was superb at doing all the physical work; however, when one of the patrons asked me for "water" and I did not understand what he meant, my boss had to intervene. For the rest of the day, I washed the grease off the garage floor, and the next day, was fired. In my defense, it was not entirely my fault—the guy had an American accent, you know?

A few days later, I started working as a dish washer in a busy restaurant called "Copper Skillet" on Sunset Boulevard. I was thankful I didn't need to talk to anyone but myself while tackling the mound of dirty dishes coming to me in trays.

Since those days, I have chosen my path leading to the mold becoming what I am today. Now, I am not a great leader, but I don't consider myself a failure, either. Nowadays, when I look back, I wish I would have worked and studied harder because we only live once, and we must make the best of it for ourselves and generations to come.

That said, I would like to share one of my secrets

with you. Every time I was frustrated or angry or felt like I wanted to quit or was dead tired while sitting in a classroom, unable to keep my eyes open, I would shake my head and say to myself, "Hey! Don't take this once-in-a-lifetime opportunity for granted! There are millions, if not billions, of people all around the world standing in a long line, wishing to be right here where you are!" And believe me, I found myself in those situations many times, and right before the tenth count, I would rise with my guard up again!

My hope is that this book helps you in whatever path you choose.

ACKNOWLEDGEMENTS

I consider myself a lucky man having such a wonderful and supportive family.

My son David, and my son-in-law, Edwin, supported me with the computer work, while my daughter Natalie and my wife Elizabeth were the best cheerleaderes to get me through the endless hours of hard work and, sometimes, self doubt.

My grandkids—Ava is six years old and her twin brothers Asher and Axel are three years old—were the best source of entertainment for me and themselves as well. All three of them would attack me to snatch my research papers, running around the house, crubmling them while screaming and laughing, and I would run after them like Charlie Chaplin just to make them have more fun.

Now, I miss that the most!

Erudition

ABOUT THE PUBLISHER

Di Angelo Publications was founded in 2008 by Sequoia Schmidt—at the age of seventeen. The modernized publishing firm's creative headquarters is in Houston, Texas, with its distribution center located in Twin Falls, Idaho. The subsidiary rights department is based in Los Angeles, and Di Angelo Publications has recently grown to include branches in England, Australia, and Sequoia's home country of New Zealand. In 2020, Di Angelo Publications made a conscious decision to move all printing and production for domestic distribution of its books to the United States. The firm is comprised of twelve imprints, and the featured imprint, Erudition, was inspired by the desire to spread knowledge, spark curiosity, and add numbers to the ranks of continuing learners, big and small.